Complex Digital Circuits

Jean-Pierre Deschamps ·
Elena Valderrama · Lluís Terés

Complex Digital Circuits

Jean-Pierre Deschamps
Tarragona, Spain

Lluís Terés
Institute of Microelectronics
of Barcelona
Campus de la UAB
Bellaterra, Spain

Elena Valderrama
Escola d'Enginyeria
Campus de la UAB
Bellaterra, Spain

ISBN 978-3-030-12655-1   ISBN 978-3-030-12653-7   (eBook)
https://doi.org/10.1007/978-3-030-12653-7

Library of Congress Control Number: 2019931840

This Springer imprint is published by the registered company Springer Nature Switzerland AG
The registered company address is: Gewerbestrasse 11, 6330 Cham, Switzerland

To our friend and colleague Jordi Aguiló, for his essential contribution to Microelectronics Research and Education.

Preface

Digital systems constitute a basic technical discipline, essential to practically any engineer. For that reason, the Engineering School of the Autonomous University of Barcelona (UAB) has designed, a couple of years ago, an introductory course entitled "Digital Systems: from Logic Gates to Processors." It is available on the Coursera massive open online course (MOOC) platform. A book including all the course material has recently been published.[1] This second book aims at continuing and at going deeper into some of the topics dealt with in the above-mentioned course and related book. So, this is not an introductory course but a more in deep approach to digital systems.

Complex systems are made up of processors executing programs, memories that store instructions and data, buses that transmit data, input–output interfaces that permit to communicate with other systems or with human users and other peripherals of different types. Many of those components are already available under the form of commercial off-the-shell products or of intellectual property (IP) cores. The latter virtual components are synthesizable descriptions in some hardware description language or even physical descriptions, for example, integrated circuit layouts for ASICs or bit-streams for FPGAs.

Thus, the development of a complex digital system generally consists in choosing components that permit to implement the desired functions and to reach the specified performance. Those components must be integrated and interconnected within some physical support (printed circuit board, multi-chip module, application-Specific integrated circuit and field-programmable gate array). Furthermore, some of those components must be programmed. Actually, a common system structure is a (set of) microprocessor(s) executing the system tasks plus several peripherals such as input–output interfaces, device drivers and others.

Some systems must also include specific (non-preexisting) components that implement algorithms whose execution on an instruction set processor should be too slow. Typical examples of such complex algorithms are: long-operand arithmetic operations, floating-point operations, encoding and processing of different types of signals, data ciphering and many others. Thus, the initial specification of the components that need a specific development work most often is an algorithm.

---

[1]Deschamps JP, Valderrama E, Terés Ll (2017) Digital Systems: from Logic Gates to Processors. Springer, New York.

The central topic of this book is the description of synthesis methods that permit to transform an initial algorithm into a specific component—a digital circuit—that satisfies some constraints such as minimum speed, maximum cost, maximum size, maximum power consumption or maximum time to market. This book is not about the development of complete and complex digital systems, a topic that includes both software and hardware aspects, but about the design of digital circuits.

Nowadays, several commercial synthesis tools permit to translate an algorithmic initial description to a digital circuit. In fact, those tools allow synthesizing the circuit in a partially automatic way: the designer generates the initial functional definition, for example, a C program, and guides the synthesis tool all along the processing steps. So, this book addresses to several types of research and development engineers. It describes synthesis methods and optimization tools, so that it addresses to developers of synthesis tools. It also addresses to developers of specific digital components, even if they use automatic synthesis tools, helping them to understand the way those tools are working and which are the choices to be made at each synthesis step.

As already pointed out, this is not an introductory text so that some previous knowledge of digital circuit design is assumed. A basic knowledge of the hardware description language VHDL is also recommended. This language is used to model digital circuits and is the input language to simulation and synthesis tools. Algorithms are defined using a pseudocode similar to VHDL. In some cases, executable VHDL processes are also used to check the correction of the proposed algorithms. All executable programs are available at the Authors' web sites www.arithmetic-circuits.org and www.cnm.es/~icas/books-courses.

Tarragona, Spain   Jean-Pierre Deschamps
Bellaterra, Spain   Elena Valderrama
Bellaterra, Spain   Lluís Terés

Acknowledgements

The authors thank the people who have helped them in developing this book, especially Prof. Gustavo Sutter who developed part of the executable programs available at the Authors' web sites. The authors are grateful to the following institutions for providing them the means for carrying this work through to a successful conclusion: Autonomous University of Barcelona (UAB) and National Center of Microelectronics (IMB-CNM-CSIC).

Contents

About the Authors

Jean-Pierre Deschamps received an MS degree in electrical engineering from the University of Louvain, Belgium, in 1967, the Ph.D. in computer science from the Autonomous University of Barcelona, Spain, in 1983, and a Ph.D. degree in electrical engineering from the Polytechnic School of Lausanne, Switzerland, in 1984. He worked in several companies and universities. His research interests include ASIC and FPGA design, and digital arithmetic. He is the author of eleven books and more than a hundred international papers.

Elena Valderrama was born in Barcelona, Catalonia, Spain. She holds a Ph.D. in Physics Science and has a degree in Medicine from the Autonoma University of Barcelona (UAB), Spain. She is currently Professor at the Microelectronics department of the UAB Engineering School. From 1980 to 1998, she was Adscript Researcher of the National Centre for Microelectronics, an institute of the Spanish Superior Board for Scientific Research (CSIC), where she led several projects in which the design and integration of highly complex digital systems (VLSI) were crucial. Her current interests focus primarily on Education, not only from the side of the professor but also on the management and quality control in Engineering-related educational programs. Her research interests move around the biomedical applications of microelectronics.

Lluís Terés received an MS degree in 1982 and the Ph.D. in 1986, both in Computer Sciences, from the Autonomous University of Barcelona (UAB). He is working in UAB since 1982 and in IMB-CNM (CSIC) since its creation in 1985. He is Head of Integrated Circuits and Systems (ICAS) group at IMB with research activity in the fields of ASIC's, sensor signal interfaces, body-implantable monitoring systems, integrated N/MEMS interfaces, flexible platform-based systems and SoC and organic/printed microelectronics. He has participated in more than seventy industrial and research projects. He is co-author of more than eighty papers and eight patents. He has participated in two spin-offs. He is also a part-time Assistant Professor at UAB.

Overview

Chapter 1 defines the classical partition of a digital circuit into data path and control unit. It starts with an introductory example. Then some general considerations are presented.

Scheduling and resource assignment are the topics of Chap. 2. In particular, the concept of precedence graph is introduced, different related optimization problems are studied, and several examples are presented.

Chapter 3 is dedicated to pipelined circuits. The main topics are circuit segmentation, combinational circuit to pipelined circuit transformation and interconnection of pipelined circuits and self-timed circuits.

The optimal implementation of loops is a basic aspect of the synthesis of digital circuits. It is the topic of Chap. 4. Combinational and sequential implementations are considered. This chapter also includes the description of techniques such as loop-unrolling and digit-serial processing.

Other topics of data path synthesis are treated in Chap. 5, for example, data path connectivity (buses), first-in first-out (FIFO) files, register files, arithmetic and logic unit (ALU), hierarchical description and sequential implementation (lower cost and longer time).

Chapter 6 is dedicated to control units. Some of the studied aspects are command encoding, hierarchical control, variable-latency operations, sequencers and microprograms.

Several examples of input–output management protocols with the corresponding interface circuits are described in Chap. 7.

The last chapter is a description of currently existing development tools, among others high-level synthesis (HLS), logic synthesis, functional simulation, logic simulation, timing analysis, intellectual property (IP) cores, formal verification, emulators and accelerators.

Architecture of Digital Circuits

The first two chapters of this book describe the classical architecture of many digital circuits and present the conventional techniques that digital circuit designers can use to translate an initial algorithmic description to an actual circuit. The main topics are the decomposition of a circuit into data path and control unit and the solution of two related problems, namely scheduling and resource assignment.

In fact, modern Electronic Design Automation tools have the capacity to directly generate circuits from algorithmic descriptions, with performances—latency, cost, consumption—comparable with those obtained with more traditional methods. Some of those development tools are described in Chap. 8.

An example of decomposition into data path and control unit is described in Sect. 4.9.1 of Deschamps et al. (2017). In this chapter, another introductory example is studied, and some general conclusions about the circuit structure and about the operation timing are presented.

1.1 Introductory Example

As a first example, a simple method for computing the base-2 logarithm of a real number is considered. Given an n-bit normalized fractional number $x = 1.x_{-1}x_{-2} \cdots x_{-n}$, compute $y = log_2x$ with an accuracy of p fractional bits. As x belongs to the interval $1 \leq x < 2$, its base-2 logarithm is a nonnegative number smaller than 1, so $y = 0.y_{-1}y_{-2} \cdots y_{-p}$.

If $y = log_2x$, then $x = 2^{0.y_{-1}y_{-2} \cdots y_{-p} \cdots}$, so that $x^2 = 2^{y_{-1}y_{-2} \cdots y_{-p} \cdots}$. Thus

- if $x^2 \geq 2$: $y_{-1} = 1$, $x' = x^2/2 = 2^{0.y_{-2} \cdots y_{-p} \cdots}$;
- if $x^2 < 2$: $y_{-1} = 0$, $x' = x^2 = 2^{0.y_{-2} \cdots y_{-p} \cdots}$.

In both cases, $x' = 2^{0.y_{-2} \cdots y_{-p} \cdots}$ so that the same method can be used to compute the value of y_{-2} and so on. The following algorithm computes y:

Algorithm 1.1 Base-2 logarithm

```
z = x; i = p;
while i > 0 loop
  z = z²;
  if z ≥ 2 then yi-p-1 = 1; z = z/2;
  else yi-p-1 = 0;
  end if;
  i = i-1;
end loop;
```

In order to check the correction of the preceding algorithm, a functional VHDL model *logarithm.vhd* has been generated and simulated. It is available at the Authors' web site. As an example, with $x = 1.691$ and $p = 16$, the result is $y = 0.1100001000000100$ (binary) $= 49{,}668/2^{16} = 0.75787353515625$ (decimal) while the value of $ln (1.691)/ln (2)$ computed with a calculator is $0.75787665974789 \cdots$. The difference is smaller than $4 \cdot 10^{-6} < 2^{-16} = 0.0000152587890625$.

To define a circuit able to execute the preceding algorithm, the following components are necessary:

- registers that store the algorithm variables,
- computation resources that execute operations,
- connections that transfer data between registers and computation resources.

A previous essential point is the definition of the data types. Algorithm 1.1 processes real numbers z. The initial value of z is a fixed-point number $z = 1.x_{-1} x_{-2} \ldots x_{-n}$. Assume that all along the computation z is represented as a fixed-point number with m fractional bits, where $m > n$. On the other hand, the value of z is always smaller than 4. Thus, z is an $(m + 2)$-bit number $z_1 z_0. z_{-1} z_{-2} \ldots z_{-m}$. Initially $z = 01. x_{-1} x_{-2} \ldots x_{-n} 0 0 \ldots 0$. The square of z is smaller than 4; it is a $(2m + 2)$-bit number, say $w_1 w_0. w_{-1} w_{-2} \ldots w_{-2m}$, that must be truncated so that $z^2 \cong w_1 w_0. w_{-1} w_{-2} \ldots w_{-m}$. If the result $z_1 z_0. z_{-1} z_{-2} \ldots z_{-m}$ of the squaring instruction [z = truncated (z^2)] is greater than or equal to 2, that is if $z_1 = 1$, then $z/2 = z_1. z_0 z_{-1} z_{-2} \ldots z_{-m}$ and this result must also be truncated so that $z/2 \cong 0 z_{1}. z_0 z_{-1} z_{-2} \ldots z_{-(m-1)}$. If the result $z_1 z_0. z_{-1} z_{-2} \ldots z_{-m}$ of the squaring instruction is smaller than 2, that is if $z_1 = 0$, then the result remains unchanged.

In conclusion, a previous mathematical (and not so easy) analysis should be necessary to define the value of m (number of fractional bits of the processed data) such that the error $|log_2 x - y|$ is smaller than 2^{-p} in spite of the rounding (truncation) operations.

Taking into account the chosen data types, Algorithm 1.1 is modified:

Algorithm 1.2 Base-2 logarithm with fixed-point data

Algorithm 1.2 processes three variables: z, y and i where

- z is an $(m + 2)$-bit fixed-point number $z_1 z_0. z_{-1} z_{-2} \ldots z_{-m}$;
- y is a p-bit fixed-point number $0.y_{-1} y_{-2} \ldots y_{-p}$;
- i is a k-bit natural such that $2^k > p$.

The circuit must contain three registers: z ($m + 2$ bits), y (p bits) and i (k bits). Registers z and i are parallel registers, but register y must permit to individually store bits y_{-1}, y_{-2} and so on, in order to execute $y_{i-p+1} = z_1$ for all values of i. A straightforward option is a set of 1-bit registers controlled by an address decoder that associates a particular enable signal to each value of i. A best option is a left shift register.

The algorithm executes three arithmetic operations, z^2, $z/2$ and $i-1$, and uses a binary condition $i > 0$. The following computation resources execute those operations:

- an $(m + 1)$-bit squaring circuit that computes $(z_0. z_{-1} z_{-2} \ldots z_{-m})^2 = w_1 w_0. w_{-1} w_{-2} \ldots w_{-2m}$;
- a divider by 2 that amounts to simple connections: if $z = z_1 z_0. z_{-1} z_{-2} \ldots z_{-m}$, then $z/2 = z_1. z_0 z_{-1} z_{-2} \ldots z_{-m}$;
- a k-bit subtractor that computes $i-1$;
- a combinational circuit that generates a binary output equal to 1 if, and only if, $i > 0$, that is the OR function of the k bits that represent i.

The corresponding circuit is shown in Fig. 1.1 (*clk* and *reset* signals are not represented). It is a *data path*, that is to say a data processor dedicated to the execution of a particular program, namely Algorithm 1.2.

In order to control the execution of the program, an additional circuit is necessary: it generates the control signals to be sent to the data path (*sel_z*, *load_z*, *sel_i*, *load_i*, *shift_y*) in

```
z₁ z₀. z₋₁z₋₂...z₋ₘ = 01.x₋₁x₋₂.x₋ₙ00...0 ; i = p;
while i > 0 loop

    z₁z₀.z₋₁z₋₂...z₋ₘ  = truncated((z₀.z₋₁z₋₂...z₋ₘ)²) ;

    yᵢ₋ₚ₋₁ = z₁ ;
    if z₁ = 1 then z = 0z₁.z₀z₋₁z₋₂...z₋₍ₘ₋₁₎ ; end if;
    i = i-1;
end loop;
```

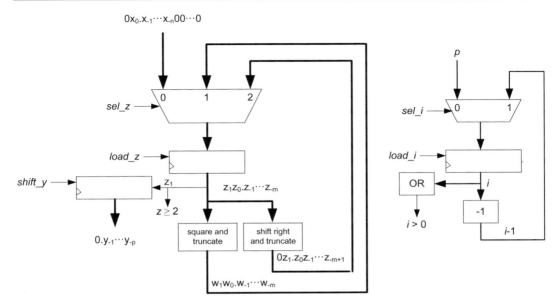

Fig. 1.1 Base-2 logarithm: data path

Fig. 1.2 Base-2 logarithm: control unit

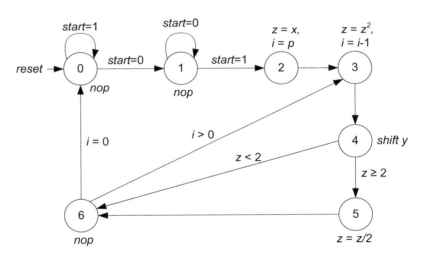

function of two Boolean conditions (flags) $i > 0$ and $z_1 = 1$. It can be modeled by the finite-state machine of Fig. 1.2.

A structural VHDL model *logarithm_circuit.vhd* that corresponds to Figs. 1.1 and 1.2 has been generated and simulated. It is available at the Authors' web site. It includes a simple communication protocol (Fig. 1.3) based on two signals: an input signal *start* and an output signal *done*: the computation starts on a positive edge of

start; then the *done* signal is lowered; it will be raised when the computation is completed.

A conclusion of this first example is that, in the case of digital circuits whose specification is an algorithm, there exists a quite natural partition of the system into two subcircuits:

- a data path that contains all resources (memory, computation and connection) necessary to execute the algorithm—a kind of specific

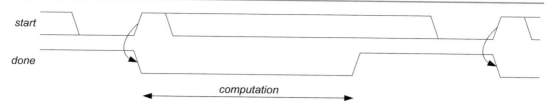

Fig. 1.3 Communication protocol

processor able to execute this particular algorithm.

- a control unit, generally modeled by a finite-state machine, that implements the program executed by the specific processor: it generates control signals such as control inputs of registers (*load*, *shift*, …), control inputs of multiplexers, programming inputs of arithmetic resources (*add*, *subtract*, …) and so on; it receives from the data path status information about, for example, the current value of some variables: zero or positive, greater than some value, negative or non-negative and so on; another example of status information could be an overflow or a by-zero division within an arithmetic block.

Obviously, the solution is not unique. The definition of a data path and a control unit implies the resolution of several optimization problems. Some of them will be treated in the next chapters. Another point to take into account is the communication protocol that permits the system to interchange data with other systems.

1.2 Data Path and Control Unit

The general structure of a digital circuit is shown in Fig. 1.4. It consists of a *data path* and a *control unit*. The data path (leftmost part of Fig. 1.4) includes computation resources executing the algorithm operations, registers storing the algorithm variables and programmable connections (e.g., multiplexers, not represented in Fig. 1.4) between resource outputs and register inputs, and between register outputs and resource inputs. The control unit (rightmost part of Fig. 1.4) is a finite-state machine. It controls the

sequence of data path operations by means of a set of control signals (*commands*) such as clock enables of registers, programming of computation resources and multiplexers and so on. It receives from the data path some feedback information (*conditions*) corresponding to the algorithm control statements (*loop*, *if*, *case* and so on).

In fact, the data path could also be considered as being a finite-state machine. Its internal states are all the possible register contents, the next-state computation is performed by the computation resources, and the output states are all the possible values of *conditions*. Nevertheless, the number of internal states is enormous and there is generally no sense to use a finite-state machine model for the data path. Any way, it is interesting to observe that the data path of Fig. 1.4 is a Moore machine (the output state only depends on the internal state) while the control unit could be a Moore or a Mealy machine. An important point is that when two finite-state machines are interconnected, one of them must be a Moore machine in order to avoid combinational loops.

According to the chronograms of Fig. 1.4, there are two critical paths: from the data registers to the internal state register and from the data registers to the data registers through the control unit. The corresponding delays are

$$T_{data-state} = t_4 + t_1 \qquad (1.1)$$

and

$$T_{data-data} = t_4 + t_2 + t_3, \qquad (1.2)$$

where t_1 is the computation time of the next internal state, t_2 the computation time of the commands, t_3 the maximum delay of the

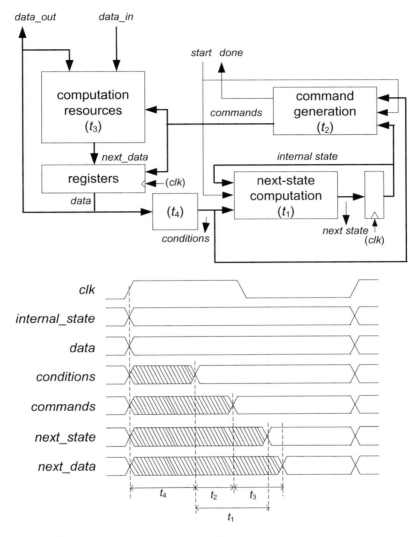

Fig. 1.4 Structure of a digital circuit: data path and control unit

computation resources and t_4 the computation time of the conditions (the setup and hold times of the registers have not been taken into account).

The clock period must satisfy

$$T_{clk} > max\{t_4 + t_1, t_4 + t_2 + t_3\}. \quad (1.3)$$

If the control unit were a Moore machine, there would be no direct path from the data registers to the data registers through the control unit, so that (1.2) and (1.3) should be replaced by

$$T_{state-data} = t_2 + t_3 \quad (1.4)$$

and

$$T_{clk} > max\{t_4 + t_1, t_2 + t_3\}. \quad (1.5)$$

Actually, it is always possible to use a Moore machine for the control unit. Generally, it has more internal states than an equivalent Mealy machine and the algorithm execution needs more clock cycles. If the values of t_1 to t_4 do not substantially vary, the conclusion could be that the Moore approach needs more, but shorter, clock cycles. Many designers also consider that Moore machines are safer than Mealy machines.

In order to increase the maximum frequency, an interesting option is to insert a command register at the output of the command generation block. Then, relation (1.2) is substituted by

$$T_{data-commands} = t_4 + t_2 \text{ and } T_{commands-data} = t_3,$$
$$(1.6)$$

so that

$$T_{clk} > max\{t_4 + t_1, t_4 + t_2, t_3\}. \qquad (1.7)$$

With this type of *registered Mealy machine*, the commands are available one cycle later than with a non-registered machine, so that additional cycles must be sometimes inserted in order that the data path and its control unit remain synchronized.

To summarize, the implementation of an algorithm is based on a decomposition of the circuit into a data path and a control unit. The data path is in charge of the algorithm operations and can be roughly defined in the following way: associate registers to the algorithm variables, implement resources able to execute the algorithm operations, and insert programmable connections (multiplexers) between the register outputs (the operands) and the resource inputs, and between the resource outputs (the results) and the register inputs. The control unit is a finite-state machine whose internal states roughly correspond to the algorithm steps, the input states are conditions (flags) generated by the data path, and the output states are commands transmitted to the data path.

In fact, the definition of a data path poses a series of optimization problems, some of them being dealt with in the next chapters, for example, scheduling of the operations, assignment of computation resources to operations and assignment of registers to variables. It is also important to notice that minor algorithm modifications sometimes yield major circuit optimizations.

1.3 Exercises

1. Appendix A briefly describes the main binary field operations. Design a circuit that computes $p(x) = a(x) \cdot b(x) \bmod f(x)$ where inputs $a(x)$ and $b(x)$ and output $p(x)$ are polynomials of degree smaller than m represented as m-bit binary vectors and constant value $f(x)$ is a polynomial of degree m. Define a data path and a control unit that execute Algorithm A1 (interleaved multiplication). Two combinational computing resources are available: the first has an m-bit input $a(x)$ and an m-bit output $y(x) = a(x) \cdot x \bmod f(x)$; the second has two m-bit inputs $a(x)$ and $c(x)$ and a 1-bit input b_i and an m-bit output $z(x) = c(x) + a(x) \cdot b_i$.

2. The following algorithm computes the greatest common divider *gcd* of two naturals a and b:

```
while a ≠ b loop
  if a < b swap(a, b); end if;
  a = a-b;
end loop;
gcd = a;
```

Define a data path and a control unit that execute the preceding algorithm and computes $z = gcd(a, b)$ where inputs a and b and output z are m-bit naturals. Three combinational computation resources are available: a magnitude comparator with two m-bit inputs a and b and two 1-bit outputs $e = 1$ if, and only if, $a = b$, and $g = 1$ if, and only if, $a > b$; a swapping circuit with two m-bit inputs a and b, a 1-bit control input c and two m-bit outputs a' and b': $(a', b') = (a, b)$ if $c = 0$, $(a', b') = (b, a)$ if $c = 1$; an m-bit subtractor.

Bibliography

Deschamps JP, Valderrama E, Terés Ll (2017) Digital Systems: from Logic Gates to Processors. Springer, New York.

Operation scheduling consists of defining which particular operations are in course of execution during every clock cycle. For that, an important concept is that of *precedence relation*. It defines which operations must be completed before starting a new one: if some result r of an operation A is an initial operand of some operation B, the computation of r must be completed before the execution of B starts. So, the execution of A must be scheduled before the execution of B.

2.1 Introductory Example

A *carry-save adder* or 3-to-2 *counter* is a circuit with 3 inputs and 2 outputs. The inputs x_i and the outputs y_j are naturals. Its behavior is defined by the following relation:

$$x_1 + x_2 + x_3 = y_1 + y_2. \qquad (2.1)$$

The basic component of a carry-save adder is a 1-bit full adder: it is a combinational circuit with three binary inputs x, y and c, and two binary outputs z and d (Fig. 2.1a). It implements the following switching functions:

$$z = x \oplus y \oplus z \text{ and } d = x \cdot y + x \cdot z + y \cdot z, \quad (2.2)$$

where $\oplus$ is the mod 2 sum (XOR function) and $+$ is the Boolean sum (OR function). In other words, the 2-bit vector (d, z) is the binary representation of the sum $x + y + z$ (in this case, the real sum):

$$x + y + z = 2 \cdot d + z. \qquad (2.3)$$

With n 1-bit full-adders, an n-bit carry-save adder can be synthesized (Fig. 2.1b with $n = 4$). In binary

$$x_1 = x_{13}x_{12}x_{11}x_{10}, x_2 = x_{23}x_{22}x_{21}x_{20},$$
$$x_3 = x_{33}x_{32}x_{31}x_{30},$$
$$y_1 = y_{13}y_{12}y_{11}y_{10}, y_2 = y_{24}y_{23}y_{22}y_{21}y_{20}.$$

According to (2.3)

$$x_{10} + x_{20} + x_{30} = 2 \cdot y_{21} + y_{10}, x_{11} + x_{21} + x_{31}$$
$$= 2 \cdot y_{22} + y_{11},$$
$$x_{12} + x_{22} + x_{32} = 2 \cdot y_{23} + y_{12}, x_{13} + x_{23} + x_{33}$$
$$= 2 \cdot y_{24} + y_{13}.$$

Then, multiply the second equation by 2, the third by 4, the fourth by 8, and add up the four equations. The result is

$$(8 \cdot x_{13} + 4 \cdot x_{12} + 2 \cdot x_{11} + x_{10})$$
$$+ (8 \cdot x_{23} + 4 \cdot x_{22} + 2 \cdot x_{21} + x_{20})$$
$$+ (8 \cdot x_{33} + 4 \cdot x_{32} + 2 \cdot x_{31} + x_{30})$$
$$= (8 \cdot y_{13} + 4 \cdot y_{12} + 2 \cdot y_{11} + y_{10})$$
$$+ (16 \cdot y_{24} + 8 \cdot y_{23} + 4 \cdot y_{22} + 2 \cdot y_{21})$$

that is (2.1).

The four (more generally n) components of the circuit of Fig. 2.1b work in parallel so that the delay of a carry-save adder is equal to the delay

© Springer Nature Switzerland AG 2019
J.-P. Deschamps et al., *Complex Digital Circuits*,
https://doi.org/10.1007/978-3-030-12653-7_2

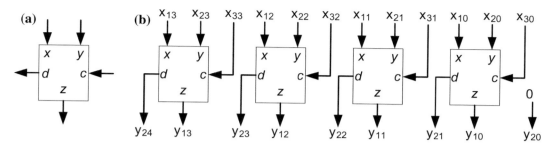

Fig. 2.1 a 1-bit full adder. **b** 4-bit carry-save adder

T_{FA} of a 1-bit full adder, independently of the number of bits of the operands.

Let CSA be the function associated with (2.1), that is

$$(y_1, y_2) = \text{CSA}(x_1, x_2, x_3). \qquad (2.4)$$

Using carry-save adders as computation resources, a 7-to-3 counter can be implemented. It allows expressing the sum of seven naturals under the form of the sum of three naturals, that is

$$x_1 + x_2 + x_3 + x_4 + x_5 + x_6 + x_7 = y_1 + y_2 + y_3.$$

In order to compute y_1, y_2 and y_3, the following operations are executed (op_1 to op_4 are labels):

$$
\begin{aligned}
op_1 &: (a_1, a_2) = \text{CSA}(x_1, x_2, x_3), \\
op_2 &: (b_1, b_2) = \text{CSA}(x_4, x_5, x_6), \\
op_3 &: (c_1, c_2) = \text{CSA}(b_1, b_2, x_7), \\
op_4 &: (d_1, d_2) = \text{CSA}(a_1, a_2, c_1).
\end{aligned} \qquad (2.5)
$$

According to (2.5) and the definition of CSA,

$$
\begin{aligned}
a_1 + a_2 &= x_1 + x_2 + x_3, \\
b_1 + b_2 &= x_4 + x_5 + x_6, \\
c_1 + c_2 &= b_1 + b_2 + x_7, \\
d_1 + d_2 &= a_1 + a_2 + c_1,
\end{aligned}
$$

so that

$$
\begin{aligned}
c_1 + c_2 + d_1 + d_2 &= b_1 + b_2 + x_7 + a_1 + a_2 + c_1 \\
&= x_1 + x_2 + x_3 + x_4 + x_5 \\
&\quad + x_6 + x_7 + c_1.
\end{aligned}
$$

Thus

$$c_2 + d_1 + d_2 = x_1 + x_2 + x_3 + x_4 + x_5 + x_6 + x_7$$

and y_1, y_2 and y_3 can be defined as follows:

$$y_1 = d_1, y_2 = d_2, y_3 = c_2.$$

The corresponding precedence relation is defined by the graph of Fig. 2.2, according to which op_2 must be executed before op_3, and op_3 before op_4. Thus, the minimum computation time is equal to $3 \cdot T_{FA}$.

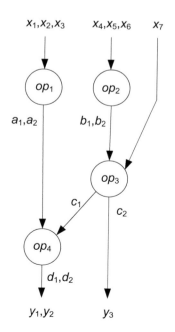

Fig. 2.2 Precedence relation of a 7-to-3 counter

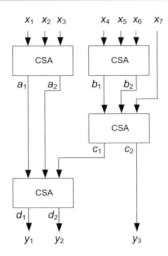

Fig. 2.3 Combinational implementation of a 7-to-3 counter

For implementing Eqs. (2.5), the following option could be considered: a combinational circuit made up of four carry-save adders (Fig. 2.3) and whose structure is the same as that of the graph of Fig. 2.2. Its computation time is equal to $3 \cdot T_{FA}$ and its cost to $4 \cdot C_{CSA}$, being C_{CSA} the cost of a carry-save adder. This is probably a bad solution because the cost is high (4 carry-save adders) and the delay is long (3 full-adders) so that the minimum clock cycle of a synchronous circuit including this 7-to-3 counter should be greater than $3 \cdot T_{FA}$.

Other options could be considered. For example, a data path including two carry-save adders and several registers (Fig. 2.4). The computation is executed in three steps:

```
0: (a₁, a₂) = CSA(x₁, x₂, x₃),
   (b₁, b₂) = CSA(x₄, x₅, x₆);
1: (c₁, c₂) = CSA(b₁, b₂, x₇);
2: (y₁, y₂) = CSA(a₁, a₂, c₁), y₂ = c₂;
```

The leftmost register stores a_1 and a_2 during cycle 0. The rightmost register stores b_1 and b_2 during cycle 0 and c_1 and c_2 during cycle 1. Observe that in this example, y_3 is a registered output while y_1 and y_2 are not. The computation time is equal to $2 \cdot T_{clk} + T_{CSA}$, where $T_{clk} > T_{CSA} = T_{FA}$, so that within a completely synchronous circuit the computation time is equal to $3 \cdot T_{clk}$, and the cost is equal to $2 \cdot C_{CSA}$, plus the cost of two registers, six 2-to-1 multiplexers and a control unit.

A third option is a data path including one carry-save adder and several registers (Fig. 2.5). The computation is executed in four cycles:

```
0: (a₁, a₂) = CSA(x₁, x₂, x₃);
1: (b₁, b₂) = CSA(x₄, x₅, x₆);
2: (c₁, y₃) = CSA(b₁, b₂, x₇);
3: (y₁, y₂) = CSA(a₁, a₂, c₁);
```

The leftmost register stores a_1 and a_2 during cycle 0 and y_1 and y_2 during cycle 3. The rightmost register stores b_1 and b_2 during cycle 1 and c_1 and y_3 during cycle 2. Observe that in this example y_1, y_2 and y_3 are registered outputs. The computation time is equal to $4 \cdot T_{clk}$, where $T_{clk} > T_{FA}$, and the cost equal to C_{CSA}, plus the cost of two registers, three 4-to-1 multiplexers and a control unit.

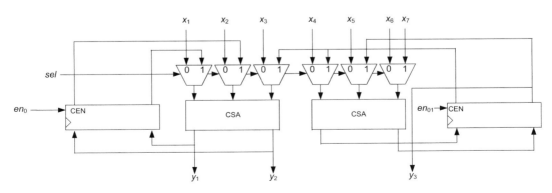

Fig. 2.4 3-cycle implementation of a 7-to-3 counter

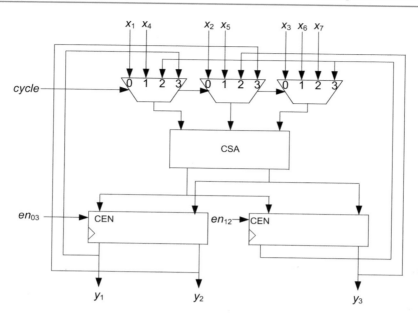

Fig. 2.5 4-cycle implementation of a 7-to-3 counter

In conclusion, to the set of operations (2.5) correspond several implementations with different costs and delays. In order to get an optimized circuit, according to some predefined criteria, the space of possible implementations must be explored. For that, optimization methods must be used.

2.2 Precedence Graph

Consider a *computation scheme*, that is to say an algorithm without branches and loops. Formally, it can be defined by a set of operations

$$op_J : (x_i, x_k, \ldots) = f(x_l, x_m, \ldots), \qquad (2.6)$$

where $x_i, x_k, x_l, x_m, \ldots$ are variables of the algorithm and f one of the algorithm operation types (*computation primitives*). Then, the precedence graph (or *data flow graph*) is defined as follows:

- Associate a vertex to each operation op_J.
- Draw an arc between vertices op_J and op_M, if one of the results generated by op_J is used by op_M.

An example was given in Sect. 2.1 (operations (2.5) and Fig. 2.2).

Assume that the computation times of all operations are known. Let t_{JM} be the computation time, expressed in number of clock cycles, of the result(s) generated by op_J and used by op_M. Then, a schedule of the algorithm is an application *Sch* from the set of vertices to the set of naturals that defines the number $Sch(op_J)$ of the cycle at the beginning of which the computation of op_J starts. A necessary condition is that

$$Sch(op_M) \geq Sch(op_J) + t_{JM} \qquad (2.7)$$

if there is an arc from op_J to op_M.

As an example, if the clock period is greater than the delay of a full adder, then, in the computation scheme (2.5), all the delays are equal to 1 and two admissible schedules are

$$Sch(op_1) = 1, Sch(op_2) = 1, Sch(op_3)$$
$$= 2, Sch(op_4) = 3, \qquad (2.8)$$

$$Sch(op_1) = 1, Sch(op_2) = 2, Sch(op_3)$$
$$= 3, Sch(op_4) = 4. \qquad (2.9)$$

Fig. 2.6 7-to-3 counter:
a ASAP schedule, **b** ALAP
schedule, **c** admissible
schedule

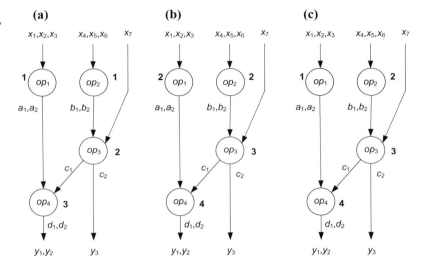

They correspond to the circuits of Figs. 2.4
and 2.5.

The definition of an admissible schedule is an
easy task. As an example, the following algorithm defines an ASAP (as soon as possible)
schedule:

- initial step: $Sch(op_J) = 1$ for all initial (without
  antecessor) vertices op_J;
- step number $n + 1$: choose an unscheduled
  vertex op_M, whose all antecessors, say op_P,
  op_Q, ... have already been scheduled, and
  define $Sch(op_M) = maximum\{Sch(op_P) + t_{PM}, Sch(op_Q) + t_{QM}, ...\}$.

Applied to (2.5), the ASAP algorithm gives
(2.8). The corresponding data flow graph is
shown in Fig. 2.6a.

An ALAP (as late as possible) schedule can
also be defined. For that, assume that the latest
admissible starting cycle for all the final vertices
(without successor) has been previously
specified:

- initial step: $Sch(op_M) =$ latest admissible
  starting cycle of op_M for all final vertices op_M;
- step number $n + 1$: choose an unscheduled
  vertex op_J, whose all successors, say op_P,
  op_Q, ... have already been scheduled, and

define $Sch(op_J) = minimum\{Sch(op_P) - t_{JP}, Sch(op_Q) - t_{JQ}, ...\}$.

Applied to (2.5), with $Sch(op_4) = 4$, the ALAP
algorithm generates

$$Sch(op_1) = 2, Sch(op_2) = 2, Sch(op_3) = 3, Sch(op_4) = 4. \quad (2.10)$$

The corresponding data flow graph is shown
in Fig. 2.6b.

Let ASAP_Sch and ALAP_Sch be ASAP and
ALAP schedules, respectively. Obviously, if op_M
is a final operation, the previously specified value
ALAP_Sch(op_M) must be greater than or equal to
ASAP_Sch(op_M). More generally, assuming that
the latest admissible starting cycle for all the final
operations has been previously specified, for any
admissible schedule Sch the following relation
holds:

$$ASAP\_Sch(op_J) \leq Sch(op_J) \leq ALAP\_Sch(op_J), \quad \forall op_J. \quad (2.11)$$

Along with (2.7), relation (2.11) defines the
admissible schedules.

An example of admissible schedule is defined
by (2.9), to which corresponds the data flow
graph of Fig. 2.9c.

A second, more realistic, example is now presented.

Example 2.1 This example corresponds to part of an Elliptic Curve Cryptography algorithm, namely the Montgomery point multiplication (Hankerson et al. 2004, Algorithm 3.40). All processed data are binary polynomials of degree smaller than some previously defined constant m:

$$a(z) = a_{m-1}z^{m-1} + a_{m-2}z^{m-2} + \ldots + a_1 z + a_0,$$
$$a_i \in \{0, 1\} \quad \forall i = 0 \text{ to } m - 1.$$

The operations used in the algorithm are the addition $a + b$ and the product $a \cdot b$ of polynomials. The product is computed modulo a polynomial

$$f(z) = z^m + f_{m-1}z^{m-1} + f_{m-2}z^{m-2} + \ldots + f_1 z + 1$$

of degree m so that $a \cdot b$ is a polynomial of degree smaller than m (Appendix A).

The following algorithm (López and Dahab 1999) computes two polynomials x_Q and y_Q in function of two polynomials x_P and y_P and of an m-bit natural $k = (k_{m-1}, k_{m-2}, \ldots, k_0)$. It executes the so-called point multiplication $(x_Q, y_Q) = k \cdot (x_P, y_P)$ for non-supersingular elliptic curves over binary field—the basic operation of several cryptographic protocols. The final *last_step* procedure includes division operations but its implementation will not be considered in this example. Observe that this procedure is executed only once, outside the main loop body, and should not significantly increase the total computation time.

Algorithm 2.1 Montgomery Point Multiplication

```
xA = 1; zA = 0; xB = xP; zB = 1;
for i in 1 .. m loop
  if k_{m-i} = 0 then
    T = zB;
    zB = (xA·zB + xB·zA)^2;
    xB = xP·zB + xA·xB·zA·T;
    T = xA;
    xA = xA^4 + zA^4;
    zA = T^2·zA^2;
  else
```

```
    T = zA;
    zA = (xA·zB + xB·zA)^2;
    xA = xP·zA + xA·xB·zB·T;
    T = xB;
    xB = xB^4 + zB^4;
    zB = T^2·zB^2;
  endif;
endloop;
(xQ, yQ) = last_step(xA, xB, zA, zB, xP, yP);
```

To implement the preceding algorithm, without the final procedure call, the following computation resources (predefined components) are available:

- adders that compute the sum $a + b$ of two polynomials in one clock cycle;
- mod f multipliers that compute the product $a \cdot b$ mod f of two polynomials in M clock cycles, where $M \geq m$; the product can be executed in M cycles, but some additional cycles are necessary to start the computation and to read the result;
- mod f squaring circuits that compute the square a^2 mod f of a polynomial in one clock cycle.

The implementation of those components is briefly described in Appendix A.

A first step of the implementation work is the modification of Algorithm 2.1 so that only 2-operand (addition and multiplication) and 1-operand (squaring) operations are used:

Algorithm 2.2 Montgomery point multiplication, version 2

```
xA = 1; zA = 0; xB = xP; zB = 1;
for i in 1 .. m loop
  if k_{m-i} = 0 then
    a = xA·zB; b = xB·zA; c = a+b; d = c^2;
    e = xP·d; f = a·b; g = e+f; h = xA·zA;
    i = h^2; j = xA+zA; k = j^2; l = k^2;
    xA = l; zA = i; xB = g; zB = d;
  else
    a = xB·zA; b = xA·zB; c = a+b; d = c^2;
    e = xP·d; f = a·b; g = e+f; h = xB·zB;
    i = h^2; j = xB+zB; k = j^2; l = k^2;
    xB = l; zB = i; xA = g; zA = d;
  endif;
```

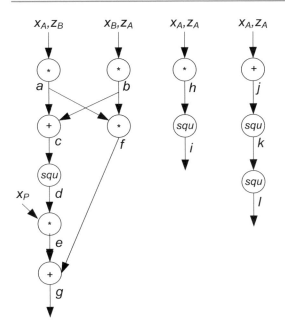

Fig. 2.7 Example 2.1: precedence graph

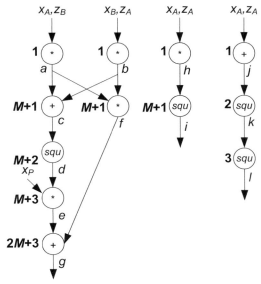

Fig. 2.8 Example 2.1: ASAP schedule

```
end loop;
(x_Q,y_Q) = last_step(x_A,x_B,z_A,z_B,x_P,y_P);
```

Equation A.1 of Appendix A has been used to compute $x_A^4 + z_A^4 = (x_A + z_A)^4$ and $x_B^4 + z_B^4 = (x_B + z_B)^4$.

Consider the main loop body of Algorithm 2.2, and assume that $k_{t-i}=0$. The corresponding sequence of operations is the computation scheme described by the data flow graph of Fig. 2.7. The operation type corresponding to every vertex is indicated (instead of operation labels). If $k_{m-i}=1$, the computation scheme is the same but for the interchange of indexes A and B.

Addition and squaring are one-cycle operations, while multiplication is an M-cycle operation. An ASAP schedule is shown in Fig. 2.8. The computation of g starts at the beginning of cycle $2M+3$, so that the final results are available at the beginning of cycle $2M+4$. The corresponding circuit must include three multipliers as the computations of a, b and h start at the same time 1.

The computation scheme includes 5 multiplications. Thus, in order to execute the algorithm with only one multiplier, the minimum number of cycles is $5M$. More precisely, one of the multiplications e,f or h cannot start before cycle $4M+1$, so that the next operation (g or i) cannot start before cycle $5M+1$. An ALAP schedule assuming that the computations of g, i and l start at the beginning of cycle $5M+1$ is shown in Fig. 2.9.

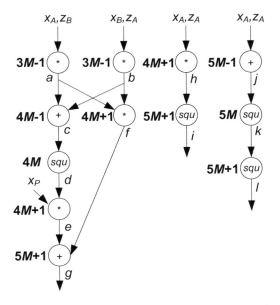

Fig. 2.9 Example 2.1: ALAP schedule with Sch $(g)=5M+1$

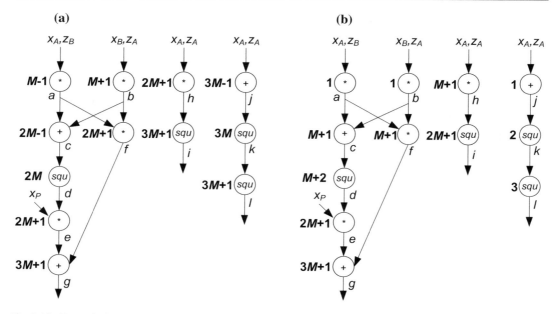

Fig. 2.10 Example 2.1: **a** ALAP schedule with $\mathrm{Sch}(g) = 3M + 1$, **b** admissible schedule

In Fig. 2.10a, an ALAP schedule with $\mathrm{Sch}(g,$ $i,$ $l) = 3M + 1$ is shown and in Fig. 2.10b an admissible schedule is shown.

Comment 2.1 A circuit based on the ASAP schedule of Fig. 2.8 must include at least three multipliers as three multiplications are scheduled at cycle 1 (a, b and h). On the other hand, the computation time is relatively short (about $2M$ cycles). With the schedule of Fig. 2.9, only one multiplier is necessary, but the computation time is long (about $5M$ cycles). With the schedule of Fig. 2.10a, three multiplications are scheduled at cycle $2M + 1$ (h, f and e), and the computation time is about $3M$ cycles. Finally, with the schedule of Fig. 2.10b, two multiplications are scheduled at cycle 1 (a and b), two multiplications at cycle $M + 1$ (h and f) and one at cycle $2M + 1$ (e) so that only two multipliers are necessary and the computation time is about $3M$. Thus, the choice of a schedule has a direct effect on the circuit performance. Some related optimization problems are dealt with in the next section.

2.3 Optimization Problems

Assuming that the latest admissible starting cycle for all the final operations has been previously specified, any schedule such that (2.7) are (2.11) hold true can be chosen. This poses optimization problems. For example:

1. Assuming that the maximum computation time has been previously specified, look for a schedule that minimizes the number of computation resources of each type.
2. Assuming that the number of available computation resources of each type has been previously specified, minimize the computation time.

An important concept is the *computation width* $w(f)$ with respect to the computation primitive (operation type) f. First define the *activity intervals* of f. Assume that f is the primitive corresponding to the operation op_J, that is

$$op_J:(x_i, x_k, \ldots) = f(x_l, x_m, \ldots).$$

Then

$$[Sch(op_J), Sch(op_J) + maximum\{t_{JM}\}]$$

is an activity interval of f. This means that a resource of type f must be available from the beginning of cycle $Sch(op_J)$ to the end of cycle $Sch(op_J) + t_{JM}$ for all M such that there is an arc from op_J to op_M. An incompatibility relation over the set of activity intervals of f can be defined: two intervals are incompatible if they overlap. If two intervals overlap, it is obvious that the corresponding operations cannot be executed by the same computation resource. Thus, a particular resource of type f must be associated with each activity interval of f in such a way that if two intervals overlap, then two distinct resources of the same type must be used. The minimum number of computation resources of type f is the *computation width* $w(f)$.

The following graphical method can be used for computing $w(f)$.

- Associate a vertex to every activity interval.
- Draw an edge between two vertices if the corresponding intervals overlap.
- Color the vertices in such a way that two vertices connected by an edge have different colors (a classical problem of graph theory).

Then, $w(f)$ is the number of different colors, and every color defines a particular resource assigned to all edges (activity intervals) with this color.

Example 2.2 Consider the scheduled precedence graph of Fig. 2.8. The activity intervals of the multiplication are

$$a: [1, M], b: [1, M], h: [1, M], e: [M+3, 2M+2], f: [M+1, 2M].$$

The corresponding incompatibility graph is shown in Fig. 2.11a. It can be colored with three colors (c_1, c_2 and c_3 in Fig. 2.11a). Thus, the computation width with respect to the multiplication is equal to 3.

If the scheduled precedence graph of Fig. 2.10b is considered, then the activity intervals of the multiplication are

$$a: [1, M], b: [1, M], h: [M+1, 2M], e: [2M+1, 3M], f: [M+1, 2M].$$

The corresponding incompatibility graph is shown in Fig. 2.11b. It can be colored with two colors. Thus, the computation width with respect to the multiplication is equal to 2.

Other schedules can be defined. According to (2.11) and Figs. 2.8 and 2.10a, the time intervals during which the five multiplications can start are the following:

$$a: [1, 3M-1], b: [1, 3M-1], h: [1, 4M+1],$$
$$e: [M+3, 4M+1], f: [M+1, 4M+1].$$

As an example, consider the admissible schedule of Fig. 2.12. The activity intervals of the multiplication operation are

$$a: [1, M], b: [M+1, 2M], h: [2M+1, 3M],$$
$$e: [4M+1, 5M], f: [3M+1, 4M].$$

Fig. 2.11 Computation width: graph coloring

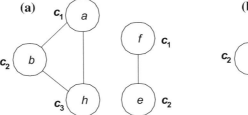

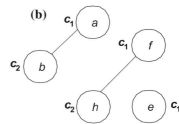

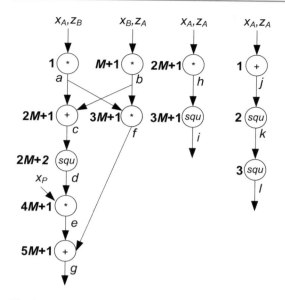

Fig. 2.12 Example 2.1: admissible schedule using only one multiplier

They do not overlap so that the incompatibility graph does not include any edge and can be colored with one color. The computation width with respect to the multiplication is equal to 1.

Thus, the two optimization problems mentioned above can be expressed in terms of computation widths:

1. Assuming that the maximum computation time has been previously specified, look for a schedule that minimizes some cost function

$$C = c_1 \cdot w(f^1) + c_2 \cdot w(f^2) + \cdots + c_m \cdot w(f^m) \tag{2.12}$$

where $f^1, f^2, \ldots, f^m$ are the computation primitives and $c_1, c_2, \ldots, c_m$ their corresponding costs.

2. Assuming that the maximum computation width $w(f)$ with respect to every computation primitive f has been previously specified, look for a schedule that minimizes the computation time.

Both are classical problems of scheduling theory. They can be expressed in terms of integer linear programming problems whose variables

are x_{It} for all operation indices I and all possible cycle numbers t: $x_{It} = 1$ if $Sch(e_I) = t$, 0 otherwise. Nevertheless, except for small computation schemes—generally tractable by hand—the so-obtained linear programs are intractable. Modern electronic design automation tools execute several types of heuristic algorithms applied to different optimization problems (not only to schedule optimization). Some of the more common heuristic strategies are *list scheduling, simulated annealing,* and *genetic algorithms.*

2.4 Resource Assignment

Once the operation schedule has been defined, several decisions must be taken.

- The number $w(f)$ of resources of type f is known, but it remains to decide which particular computation resource executes each operation. Furthermore, the definition of multifunctional programmable resources could also be considered.
- As regards the storing resources, a simple solution is to assign a particular register to every variable. Nevertheless, in some cases the same register can be used for storing different variables.

A key concept for assigning registers to variables is the *lifetime* $[t_I, t_J]$ of every variable: t_I is the number of the cycle during which its value is generated, and t_J is the number of the last cycle during which its value is used.

Example 2.3 Consider the computation scheme of Fig. 2.7 and the schedule of Fig. 2.12. The computation width is equal to 1 for all primitives (multiplication, addition and squaring).

In order to compute the variable lifetimes, it is assumed that the operations are executed as follows: in the case of an M-cycle operation such as $p = s \cdot t \bmod f$ scheduled at cycle I:

- The assigned multiplier reads the operands from the registers assigned to s and t, internally stores them and starts computing $s \cdot t$.

- The result is generated and available on the multiplier output during cycle number $I + M - 1$ (or sooner) and is stored within the register assigned to p at the end of this same cycle.
- Thus, the value of p is available for any operation scheduled at cycle number $I + M$ or later.

In the case of a one-cycle operation such as $p = s + t$ or $p = s^2 \bmod f$, scheduled at cycle I, the corresponding computing resource is a combinational circuit; the result is generated and available on this combinational circuit output during cycle number I and is stored within the register assigned to p at the end of this same cycle; thus, the value of p is available for any operation scheduled at cycle number $I + 1$ or later.

Taking into account the preceding rules, the computation is executed as follows:

number $I + M - 1 = 4M$ and is stored at the end of this cycle within the register assigned to f; the value of f is available for any operation beginning at cycle number $4M + 1$ or later, for example $g = e + f$ executed at cycle $5M + 1$.

As regards the variables x_A, z_A, x_B and z_B, in charge of passing values from one iteration step to the next one (Algorithm 2.2), their values are available from the beginning of the computation scheme execution and must remain available up to the last cycle during which those values are used. At the end of the computation scheme execution, they must be updated with their new values.

The lifetime intervals are given in Table 2.1. As an example, the value of a is generated and stored during cycle number M, and the last cycle during which the value of a is used is cycle number $3M + 1$ (when a is read and internally stored within the multiplier). Thus, the lifetime interval of a is $[M, 3M + 1]$.

```
initial data: xA, zA, xB, zB
cycle 1:   j = xA+zA; read, store and start xA·zB;
cycle 2:   k = j²;
cycle 3:   l = k²;
cycle M:   a = multiplier_output;
cycle M+1: read, store and start xB·zA;
cycle 2M:  b = multiplier_output;
cycle 2M+1: c = a + b; read, store and start xA·zA;
cycle 2M+2: d = c²;
cycle 3M:  h = multiplier_output;
cycle 3M+1: i = h²; read, store and start a·b;
cycle 4M:  f = multiplier_output;
cycle 4M+1: read, store and start xP·d;
cycle 5M:  e = multiplier_output;
cycle 5M+1: g = e+f;
final
results: (xA, zA, xB, zB) := (l, i, g, d);
```

For example, the multiplier executes $f = a \cdot b$ as follows: during cycle number $I = 3M + 1$, the multiplier reads and internally stores the values of a and b, and the multiplication execution begins; the result is generated during cycle

The definition of a minimum number of registers can be expressed as a graph coloring problem. For that, associate a vertex to every variable and draw an edge between two variables if their lifetime intervals are incompatible.

Table 2.1 Lifetime intervals

a	$[M, 3M + 1]$
b	$[2M, 3M + 1]$
c	$[2M + 1, 3M + 1]$
d	$[2M + 2, final]$
e	$[5M, 5M + 1]$
f	$[4M, 5M + 1]$
g	$[5M + 1, final]$
h	$[3M, 3M + 1]$
i	$[3M + 1, final]$
j	$[1, 2]$
k	$[2, 3]$
l	$[3, final]$
x_A	$[initial, 2M + 1]$
z_A	$[initial, 2M + 1]$
x_B	$[initial, M + 1]$
z_B	$[initial, 1]$

Definition 2.1

Two lifetime intervals are incompatible if they have more than one common cycle. Actually, lifetime intervals such as $[t, u]$ and $[u, v]$ with one common cycle (number u) are compatible: $[t, u]$ corresponds to a variable y_1 stored at the end of cycle number t and used during cycle number $u > t$, and $[u, v]$ corresponds to a variable y_2 stored at the end of cycle number u and used during cycle number $v > u$. Thus, if edge-triggered registers are used (flip-flops), the same register can store the value of y_1 all along

cycle number u and sample the value of y_2 at the end of the same cycle number u (Fig. 2.13).

As an example, the lifetime intervals $[1, 2]$ and $[2, 3]$ of j and k are compatible; they correspond to two successive instructions

```
cycle 1: j = x_A + z_A;
cycle 2: k = j²;
```

that may be substituted by

```
cycle 1: R = x_A + z_A;
cycle 2: R = R²;
```

At the end of cycle number 1, the value of $x_A + z_A$ is stored within register R. During cycle number 2, the square of R is computed, and the result is stored in the same register R at the end of the same cycle.

Similarly, the lifetime intervals $[M, 3M + 1]$ and $[3M + 1, final]$ of a and i are compatible; they correspond to instructions

```
cycle M: a = multiplier_output;
...
cycle 3M+1: i = h²; read, store and start
a·b;
```

that may be substituted by

```
cycle M: R = multiplier_output;
...
cycle 3M+1: R = h²; read, store and start
R·b;
```

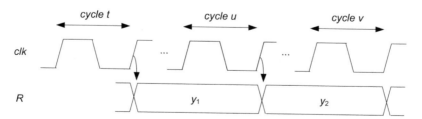

Fig. 2.13 Life intervals with a common cycle number u

At the end of cycle number M, the multiplier output value is stored within register R. During cycle number $3M + 1$, the value of R is stored within the multiplier, the square of h is computed, and the result h^2 is stored within the same register R at the end of the same cycle.

On the other hand, the lifetime intervals $[M, 3M + 1]$ and $[2M, 3M + 1]$ of a and b are not compatible as they correspond to instructions

z_B (initial $\rightarrow 1$), j $(1 \rightarrow 2)$, k $(2 \rightarrow 3)$, $l(3 \rightarrow \text{final})$; x_B (initial $\rightarrow M + 1$), b $(2M \rightarrow 3M + 1)$, f $(4M \rightarrow 5M + 1)$, $g(5M + 1 \rightarrow \text{final})$; z_A (initial $\rightarrow 2M + 1$), c $(2M + 1 \rightarrow 2M + 2)$, d $(2M + 2 \rightarrow \text{final})$; x_A (initial $\rightarrow 2M + 1$), h $(3M \rightarrow 3M + 1)$, e $(5M \rightarrow 5M + 1)$; a $(M \rightarrow 3M + 1)$, i $(3M + 1 \rightarrow \text{final})$.

Thus, the computing scheme can be executed with five registers:

```
cycle M: a = multiplier_output;
...
cycle 2M: b = multiplier_output;
...
cycle 3M+1: i = h²; read, store and start a·b;
```

The (generally different) values of a and b cannot be saved within the same register.

The following groups of variables have compatible lifetime intervals:

- x_A stores the initial value of x_A, h and e;
- z_A stores the initial value of z_A, c and d;
- x_B stores the initial value of x_B, b, f and g;
- z_B stores the initial value of z_B, j, k and l;
- R: stores a and i.

```
cycle 1:    z_B = x_A + z_A; read, store and start x_A·z_B;
cycle 2:    z_B = z_B²;
cycle 3:    z_B = z_B²;
cycle 3M:   R = multiplier_output;
cycle 3M+1: read, store and start x_B·z_A;
cycle 2M:   x_B = multiplier_output;
cycle 2M+1: z_A = R + x_B; read, store and start x_A·z_A;
cycle 2M+2: z_A = z_A²;
cycle 3M:   x_A = multiplier_output;
cycle 3M+1: R = x_A²; read, store and start start R·x_B;
cycle 4M:   x_B = multiplier_output;
cycle 4M+1: read, store and start x_P·z_A;
cycle 5M:   x_A = multiplier_output;
cycle 5M+1: x_B = x_A + x_B;
final results: (x_A, z_A, x_B, z_B) = (z_B, R, x_B, z_A);
```

Obviously, the last two cycles can be replaced by

```
cycle 5M+1: (x_A, z_A, x_B, z_B) = (z_B, R, x_A+x_B,
z_A);
```

Example 2.4 Consider the same computation scheme (Fig. 2.7) with a different schedule (Fig. 2.10b). The computation width with respect to multiplication is equal to 2 and is equal to 1 with respect to addition and squaring. Thus, two multipliers (multiplier1 and multiplier2) must be used. The computation is executed as follows:

z_B (*initial* $\rightarrow$ 1), j (1 $\rightarrow$ 2), k (2 $\rightarrow$ 3), l(3 $\rightarrow$ *final*);

x_B (*initial* $\rightarrow$ 1), b ($M \rightarrow M+1$), f ($2M \rightarrow 3M + 1$), g($3M + 1 \rightarrow$ *final*);

z_A (*initial* $\rightarrow M + 1$), c ($M + 1 \rightarrow M + 2$), d ($M + 2 \rightarrow$ *final*);

x_A (*initial* $\rightarrow M + 1$), h ($2M \rightarrow 2M + 1$), i ($2M + 1 \rightarrow$ *final*);

a ($M \rightarrow M + 1$), e ($3M \rightarrow 3M + 1$).

Thus, the computing scheme can be executed with five registers:

```
initial data: x_A, z_A, x_B, z_B
cycle 1:   j = x_A+z_A; read, store and start x_A·z_B and x_B·z_A;
cycle 2:   k = j²;
cycle 3:   l = k²;
cycle M:   a = multiplier1_output; b = multiplier2_output;
cycle M+1: c = a + b; read, store and start a·b and x_A·z_A;
cycle M+2: d = c²;
cycle 2M:  f = multiplier1_output; h = multiplier2_output;
cycle 2M+1: i = h²; read, store and start x_P·d;
cycle 3M:  e = multiplier2_output;
cycle 3M+1: g = e + f;
final results: (x_A, z_A, x_B, z_B) = (l, i, g, d);
```

The lifetime intervals are given in Table 2.2.

The following groups of variables have compatible lifetime intervals:

- x_A stores the initial value of x_A, h and i;
- z_A stores the initial value of z_A, c and d;
- x_B stores the initial value of x_B, b, f and g;
- z_B stores the initial value of z_B, j, k and l;
- R: stores a and i.

```
cycle 1:   z_B = x_A+z_A; read, store and start x_A·z_B and x_B·z_A;
cycle 2:   z_B = z_B²;
cycle 3:   z_B = z_B²;
cycle M:   R = multiplier1_output; x_B = multiplier2_output;
cycle M+1: z_A = R+x_B; read, store and start R·x_B and x_A·z_A;
cycle M+2: z_A = z_A²;
cycle 2M:  x_B = multiplier1_output; x_A = multiplier2_output;
cycle 2M+1: x_A = x_A²; read, store and start x_P·z_A;
cycle 3M:  R = multiplier1_output;
cycle 3M+1: (x_A, z_A, x_B, z_B) := (z_B, x_A, R+x_B, z_A);
```

Table 2.2 Lifetime intervals

a	$[M, M+1]$
b	$[M, M+1]$
c	$[M+1, M+2]$
d	$[M+2, final]$
e	$[3M, 3M+1]$
f	$[2M, 3M+1]$
g	$[3M+1, final]$
h	$[2M, 2M+1]$
i	$[2M+1, final]$
j	$[1, 2]$
k	$[2, 3]$
l	$[3, final]$
x_A	$[initial, M+1]$
z_A	$[initial, M+1]$
x_B	$[initial, 1]$
z_B	$[initial, 1]$

Compare Examples 2.3 and 2.4 based on the schedules of Figs. 2.12 and 1.10b, respectively. The data path that corresponds to Example 2.3 must include a multiplier, a squaring circuit, an adder and five registers, plus connection resources, for example multiplexers. It executes the computation in about $5M$ cycles. The data path that corresponds to Example 2.3 must include two multipliers, a squaring circuit, an adder and five registers, plus connection resources. It executes the computation in about $3M$ cycles. The second solution is faster ($3M$ cycles instead of $5M$) but needs two multipliers.

2.5 Final Example

To conclude this chapter, Algorithm 2.2 without the final *last_step* procedure is implemented.

2.5.1 Data Path

The loop body consists of two computation schemes, either the scheme of Fig. 2.7 executed when $k_{m-i}=0$, or a similar one when $k_{m-i}=1$. The schedule of Fig. 2.12 is chosen so that only one multiplier is necessary.

The following computation resources are necessary:

- an adder that computes the sum $A+B$ of two polynomials;
- a mod f multiplier that computes the product $A \cdot B \bmod f$ of two polynomials;
- a mod f squaring circuit that computes the square $a^2 \bmod f$ of a polynomial.

Field addition amounts to bit-by-bit modulo 2 additions (XOR functions) so that the adder is a combinational circuit consisting of m XOR gates working in parallel (Fig. 2.14a). It computes $C = A + B$ in one cycle.

Computation resources executing field squaring and multiplication have been described in (Deschamps et al. 2009). Complete and synthesizable source files *classic_squarer.vhd* and *interleaved_mult.vhd* are available at the Authors' web site. They are considered as predefined IP components (intellectual property components) whose main characteristics are the following.

- The *classic squarer* is a combinational circuit that computes $c = a^2 \bmod f$ in one cycle (Fig. 2.14c).
- The *interleaved multiplier* computes the product $Z = A \cdot B \bmod f$ of two polynomials in M cycles, with $M > m$, being m the degree of polynomial f (Fig. 2.14b). It communicates with other circuits with two signals *start* and

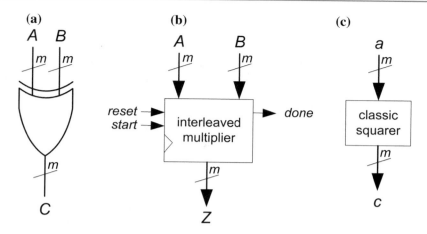

Fig. 2.14 Computation resources

done: it reads and internally stores the input operands during the first cycle after detecting a positive edge on *start* and raises an output flag *done* when the multiplication result is available (Fig. 2.15).

A number *M* of cycles has been previously used to define admissible schedules. However, as

the multiplier raises a flag *done* when the result is available, it is not necessary to use a mod *M* counter to determine when the result is available on the multiplier output. Algorithm 2.2 without the final *last_step* procedure is equivalent to the following that includes *wait* instructions. Sentences separated by commas are executed in parallel:

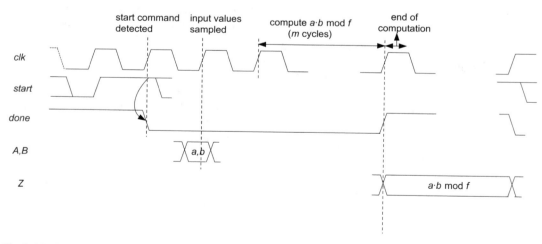

Fig. 2.15 Communication protocol

Algorithm 2.3

```
xA = 1, zA = 0, xB = xP, zB = 1;
for i in 1 .. m loop
 if k_{m-i} = 0 then
   zB = xA+zA, start (Z = xA·zB);
   zB = zB²;
   zB = zB²;
   wait until done;
   R = Z;
   start (Z = xB·zA);
   wait until done;
   xB = Z;
   zA = R + xB, start (Z = xA·zA);
   zA = zA²;
   wait until done;
   xA = Z;
   R = xA², start(Z = R·xB);
   wait until done;
   xB = Z;
   start (Z = xP·zA);
   wait until done;
   xA = Z;
   (xA, zA, xB, zB) = (zB, R, xA+xB, zA);
 else
   zA = xB+zB, start(Z = xB·zA);
   zA = zA²;
   zA = zA²;
   wait until done;
   R = Z;
   start (Z = xA·zB);
   wait until done;
   xA = Z;
   zB = R + xA, start (Z = xB·zB);
   zB = zB²;
   wait until done;
   xB = Z;
   R = xB², start(Z = R·xA);
   wait until done;
   xA = Z;
   start (Z = xP·zB);
   wait until done;
   xB = Z;
```

```
   (xB, zB, xA, zA) = (zA, R, xB+xA, zB);
 end if;
end loop;
```

A data path able to execute Algorithm 2.3 must include

- three computation resources: mod f adder, multiplier and squaring circuit;
- five registers that store m-bit data: x_A, x_B, z_A, z_B and R;
- controllable connections to transfer data between external inputs and outputs, computation resources and registers.

It must also include a shift register that permits to sequentially read the bits of k and a mod m counter to control the loop execution.

To specify the connection resources, consider first the set of products included in Algorithm 2.3. There are eight different products:

$$x_A \cdot z_B, x_B \cdot z_A, x_A \cdot z_A, R \cdot x_B, x_P \cdot z_A, x_B \cdot z_B,$$
$$R \cdot x_A, x_P \cdot z_B.$$

In order to connect the external input x_P and the five register outputs x_A, x_B, z_A, z_B and R to the two multiplier operand inputs, a straightforward solution is shown in Fig. 2.16a: two 4-to-1 m-bit multiplexers. However, as mod f product is a commutative operation, a better solution could be considered. For that define an incompatibility relation over the set $\{x_A, x_B, z_A, z_B, R, x_P\}$: two elements are incompatible if they are operands of a same operation. As an example, x_A and z_B are incompatible, as $x_A \cdot z_B$ is one of the operations. The corresponding graph (Fig. 2.16b) can be colored with two colors corresponding to the sets $\{x_A, x_B, x_P\}$ and $\{z_A, z_B, R\}$. Thus, none of the operations is a product of two elements of the same set, so that x_A, x_B and x_P can be assigned to

Fig. 2.16 Inputs to the multiplier: **a** straightforward solution, **b** colored graph

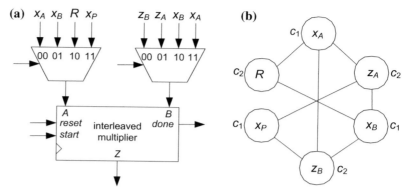

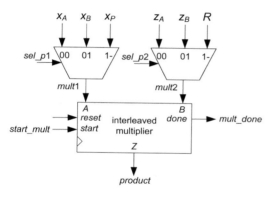

Fig. 2.17 Inputs to the multiplier: a better solution

the leftmost multiplier input and z_A, z_B and R to the rightmost input. For that, two 3-to-1 m-bit multiplexers are used (Fig. 2.17).

An important observation: the circuit of Fig. 2.17 works correctly if the multiplexer outputs $mult1$ and $mult2$ transmit the correct operand values during the first cycle after the detection of a positive edge on $start$ (Fig. 2.15). In particular, operations such as $z_B = x_A + z_A$ and $start(Z = x_A \cdot z_B)$ could not be executed in parallel because the first one modifies the value of z_B before the sampling and storing of z_B within the interleaved multiplier. A first solution is to modify Algorithm 2.3 and to replace instructions such as

$z_B = x_A + z_A$, $start(Z = x_A \cdot z_B)$;

by two successive instructions

$start(Z = x_A \cdot z_B)$;
$z_B = x_A + z_A$;

The computation time is a bit longer (actually three more cycles). Another option that permit to execute in parallel operations such as

$z_B = x_A + z_A$, $start(Z = x_A \cdot z_B)$;

is to add a $2m$-bit register (Fig. 2.18): in this way, the multiplexer output values $mult1$ and $mult2$ are sampled when $start = 1$ and the register outputs $mult1\_reg$ and $mult2\_reg$ transmit the correct operand values during the next cycle (the first cycle after the detection of a positive edge on $start$).

The same type of analysis must be done to define the connections to the adder inputs. The operations executed by the adder are

$x_A + z_A, R + x_B, x_A + x_B, x_B + z_B, R + x_A, x_B + x_A.$

In this case, the incompatibility graph must be colored with three colors (Fig. 2.19a) corresponding to the sets $\{x_A, z_B\}$, $\{x_B, z_A\}$ and $\{R\}$. A possible solution is to assign the first set to the

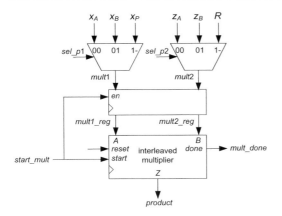

Fig. 2.18 Registered inputs to the multiplier

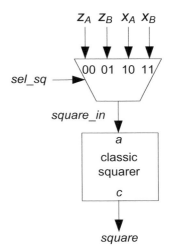

Fig. 2.20 Inputs to the squaring circuit

leftmost adder input, the second to the rightmost input, and R to both inputs. Two 3-to-1 m-bit multiplexers are used (Fig. 2.19b).

Finally, the operations realized by the squaring primitive are

$$z_B^2, z_A^2, x_A^2, x_B^2.$$

A 4-to-1 m-bit multiplexer is used (Fig. 2.20).

Consider now the storing resources. Assuming that x_P and k are input variables that remain available during the whole algorithm execution, there remain five variables that must be internally stored: x_A, x_B, z_A, z_B and R. For every register, the origin of the data stored in every register must be defined.

The operations that update x_A are

$$x_A = 1; x_A = Z; x_A = z_B; x_A = x_B + x_A;$$

So, the updated value can be: 1 (initial value), the multiplier output *product*, the adder output *adder_out* or z_B. The corresponding part of the data path is an m-bit register, that initially stores $000\cdots01$ (when *load* = 1), and a 3-to-1 m-bit multiplexer (Fig. 2.21a).

The operations that update x_B are

$$x_B = x_P; x_B = Z; x_B = x_A + x_B; x_B = z_A;$$

So, the updated value can be: x_P (initial value), the multiplier output *product*, the adder output *adder_out* or z_A. The corresponding part of the data path is an m-bit register that initially stores x_P, and a 3-to-1 m-bit multiplexer (Fig. 2.21b).

(a)   **(b)**

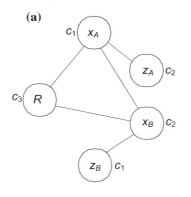

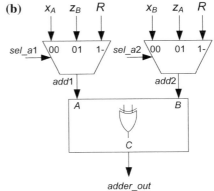

Fig. 2.19 Inputs to the adder

Fig. 2.21 Data path registers

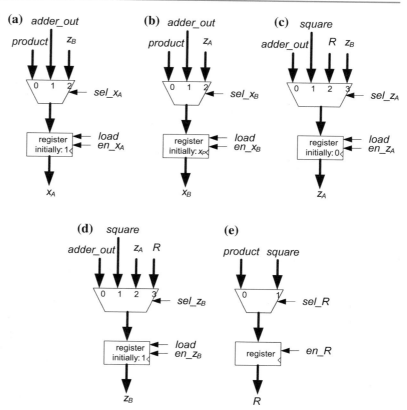

The operations that update z_A are

$$z_A = 0; z_A = R + x_B; z_A = z_A^2;$$
$$z_A = R; z_A = x_B + z_B; z_A = z_B;$$

So, the updated value can be: 0 (initial value), the adder output *adder_out*, the squaring circuit output *square*, R or z_B. The corresponding part of the data path is an m-bit register, that initially stores $000\cdots00$, and a 4-to-1 m-bit multiplexer (Fig. 2.21c).

The operations that update z_B are

$$z_B = 1; z_B = x_A + z_A; z_B = z_B^2;$$
$$z_B = z_A; z_B = R + x_A; z_B = R;$$

So, the updated value can be: 1 (initial value), the adder output *adder_out*, the squaring circuit output *square*, z_A or R. The corresponding part of the data path is an m-bit register, that initially

stores $000\cdots01$, and a 4-to-1 m-bit multiplexer (Fig. 2.21d).

Finally, the operations that update R are

$$R = Z; R = x_A^2; R = x_B^2;$$

So, the updated value can be: the multiplier output *product* or the squaring circuit output *square*. The corresponding part of the data path is an m-bit register and a 2-to-1 m-bit multiplexer (Fig. 2.21e).

The data path includes two additional components: a shift register that permits to sequentially read the bits of k and a mod m counter to control the loop execution. They are shown in Fig. 2.22. The shift register works as follows:

- When $load = 1$, the value of k (a circuit input) is stored within the internal register *internal_k*, so that *internal_k* = k.

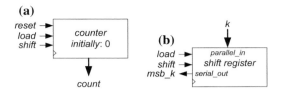

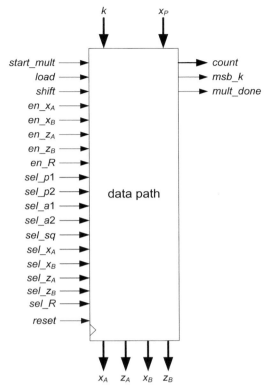

Fig. 2.22 Shift register and counter

- When $shift = 1$, the register contents are shifted one bit to the left: $internal\_k = internal\_k$ $(m - 2 \cdots 0)$ & 0.
- The output $msb\text{-}\_k = internal\_k(m - 1)$.

In this way, the m bits of k, initially stored in $internal\_k$, are outputted through the serial msb_k output; first k_{m-1}, then k_{m-2}, and so on.

The counter works as follows:

- When $load = 1$, the value of $count$ is reset to 0.
- When $shift = 1$ the counter contents are incremented: $count = count + 1$.

The complete data path (Fig. 2.23) is made up of the components of Fig. 2.18 (interleaved multiplier and multiplexers with registered outputs), Fig. 2.19b (adder and input multiplexers), Fig. 2.20 (squaring circuit and input multiplexer), Fig. 2.21 (parallel registers with their corresponding input multiplexers) and Fig. 2.22 (mod m counter and left shift register).

Fig. 2.23 Data path

A complete VHDL model *scalar_product_-data_path.vhd* is available at the Authors' web site. The numbers of bits of the input and output ports are defined in the following VHDL entity declaration.

```
ENTITY scalar_product_data_path IS
PORT (
xP, k: IN STD_LOGIC_VECTOR (m-1 DOWNTO 0);
clk, reset, start_mult, load, shift, en_XA,
en_XB, en_ZA, en_ZB, en_R: IN STD_LOGIC;
sel_p1, sel_p2, sel_a1, sel_a2, sel_sq, sel_xA,
sel_xB, sel_zA, sel_zB:
 IN STD_LOGIC_VECTOR (1 DOWNTO 0);
sel_R: IN STD_LOGIC;
xA, zA, xB, zB:
 INOUT STD_LOGIC_VECTOR (m-1 DOWNTO 0);
count: INOUT NATURAL RANGE 0 TO m-1;
msb_k, mult_done: OUT STD_LOGIC
);
END scalar_product_data_path;
```

2.5.2 Complete Circuit

The complete circuit structure is shown in Fig. 2.24. Thus, it remains to generate the control unit.

The generation of the control unit is a task similar to the translation of a programming language program to an equivalent program in the machine language corresponding to a particular processor.

- On the one hand, an algorithm has been defined (Algorithm 2.3). It can be considered as a program in a programming language—in this case pseudocode.
- On the other hand, a synthesizable VHDL model of a data path has been generated

(Fig. 2.23, *scalar_product_data_path.vhd*). It can be considered as a specific processor able to execute the pseudocode program.

The operations that the data path executes correspond to all possible value combinations of signals

- *start_mult, load, shift, en_X_A, en_X_B, en_Z_A, en_Z_B, en_R, sel_p1, sel_p2, sel_a1, sel_a2, sel_sq, sel_x_A, sel_x_B, sel_z_A, sel_z_B, sel_R,*

in total 27 bits (nine 1-bit signals and nine 2-bit signals).

The program execution control is based on the values of four signals

Fig. 2.24 Structure of the scalar product circuit

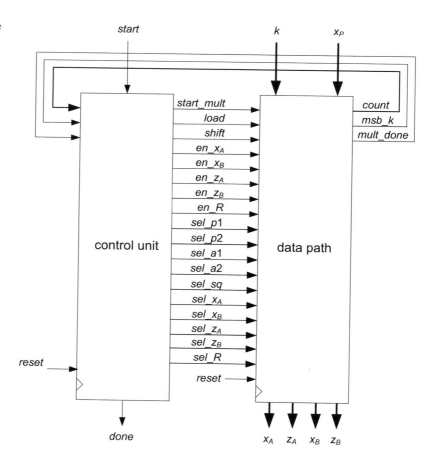

- *count*, *msb_k*, *mult_done* generated by the data path;
- *start* (an external control signal),

whose values define branching or jump conditions.

As seen before (Sect. 1.2), the control unit is modeled by a finite state machine. In this case, an easy way to specify the finite state machine is to associate an internal state with each operation, branching, or jump of Algorithm 2.3. More precisely:

- Two initial instructions are added to detect a positive edge of *start*.
- An output *done* is generated: it is equal to 0 from the beginning to the end of the computation and equal to 1 as long as a *start* order (positive edge of *start*) is waited for.
- A simple and obvious modification of Algorithm 2.3 generates the following algorithm in which only operations and *wait until* and *go to* statements are used.

Algorithm 2.4 Internal states and operations (Algorithm 2.3 modified)

```
0: wait until start = 0, done = 1;
1: wait until start = 1;
2: x_A = 1, z_A = 0, x_B = x_P, z_B = 1, count = 0,
done = 0;
3: if k_{m-i} = 1 then go to 23;
4: z_B = x_A + z_A, start (Z = x_A·z_B);
5: z_B = z_B^2;
6: z_B = z_B^2;
7: wait until mult_done = 1;
8: R = Z;
9: start (Z = x_B·z_A);
10: wait until mult_done = 1;
11: x_B = Z;
12: z_A = R + x_B, start (Z = x_A·z_A);
13: z_A = z_A^2;
14: wait until mult_done = 1;
15: x_A = Z;
16: R = x_A^2, start (Z = R·x_B);
17: wait until mult_done = 1;
18: x_B = Z;
```

```
19: start (Z = x_P·z_A);
20: wait until mult_done = 1;
21: x_A = Z;
22: (x_A, z_A, x_B, z_B) = (z_B, R, x_A+x_B, z_A), go
to 42;
23: z_A = x_B+z_B, start (Z = x_B·z_A);
24: z_A = z_A^2;
25: z_A = z_A^2;
26: wait until mult_done = 1;
27: R = Z;
28: start (Z = x_A·z_B);
29: wait until mult_done = 1;
30: x_A = Z;
31: z_B = R + x_A, start (Z = x_B·z_B);
32: z_B = z_B^2;
33: wait until mult_done = 1;
34: x_B = Z;
35: R = x_B^2, start (Z = R·x_A);
36: wait until mult_done = 1;
37: x_A = Z;
38: start (Z = x_P·z_B);
39: wait until mult_done = 1;
40: x_B = Z;
41: (x_B, z_B, x_A, z_A) = (z_A, R, x_B+x_A, z_B);
42: if count < m-1 then count = count +1,    go
to 3;
  else go to 0;
```

The preceding program is executed in a sequential manner, except when an explicit jump (go to) instruction is included. The corresponding finite state machine has 43 internal states $0, 1, 2, \cdots, 42$. The next-state function can be directly extracted from the preceding program (Algorithm 2.4). For example, if the current internal state is 0, that corresponds to instruction

```
0: wait until start = 0, done = 1;
```

then the next internal state is 1 if $start = 0$ and 0 (does not change) if $start = 1$.

If the current internal state is 2, that corresponds to instruction

```
2: x_A = 1, z_A = 0, x_B = x_P, z_B = 1, count = 0,
done = 0;
```

then the next internal state is 3.

A last example: if the current internal state is 42, that corresponds to instruction

```
42: if count < m-1 then count = count +1,
go to 3;
  else go to 0;
```

then the next internal state is 3 if $count < m - 1$ and is 0 if $count = m - 1$.

As in many cases, the next internal state is the current state plus 1; an interesting option is the use of an implicit program counter: for example, in the case of instruction number 0, the next internal state is $current\_state + 1$ if $start = 0$ and 0 (does not change) if $start = 1$. Or, in the case of instruction number 2, the next internal state is $current\_state + 1$.

Finally, the next-state function of the control finite state machine is defined by the following *case* statement.

Algorithm 2.5 Next-state function

```
case current_state is
when 0 => if start = 0 then current_state = c-
urrent_state + 1;
    end if;
when 1 => if start = 1 then
current_state = current_state + 1;
    end if;
when 2 => current_state =
current_state + 1;
when 3 => if msb_k = 0 then
current_state = 4;
    else current_state = 23; end if;
when 4 to 6 => current_state =
current_state + 1;
when 7 => if mult_done = 1 then
current_state = current_state + 1;
    end if;
when 8 to 9 => current_state =
current_state + 1;
when 10 => if mult_done = 1 then
current_state = current_state + 1;
    end if;
when 11 to 13 => current_state =
current_state + 1;
when 14 => if mult_done = 1 then
current_state = current_state + 1;
    end if;
when 15 to 16 => current_state =
current_state + 1;
when 17 => if mult_done = 1 then
current_state = current_state + 1;
    end if;
when 18 to 19 => current_state =
current_state + 1;
when 20 => if mult_done = 1 then
current_state = current_state + 1;
    end if;
when 21 => current_state =
current_state + 1;
when 22 => current_state = 42;
when 23 to 25 => current_state =
current_state + 1;
when 26 => if mult_done = 1 then
current_state = current_state + 1;
    end if;
when 27 to 28 => current_state =
current_state + 1;
when 29 => if mult_done = 1 then
current_state = current_state + 1;
    end if;
when 30 to 32 => current_state =
current_state + 1;
when 33 => if mult_done = 1 then
current_state = current_state + 1;
    end if;
when 34 to 35 => current_state =
current_state + 1;
when 36 => if mult_done = 1 then
current_state = current_state + 1;
    end if;
when 37 to 38 => current_state =
current_state + 1;
when 39 => if mult_done = 1 then
current_state = current_state + 1;
    end if;
when 40 => current_state =
current_state + 1;
when 41 => current_state =
current_state + 1;
when 42 => if count < m-1 then
current_state = 3;
    else current_state = 0; end if;
end case;
```

The definition of the output function is a conceptually easy but rather tedious task: it consists of associating with each operation—or set of operations—of Algorithm 2.4 the values of the control signals that initiate those operations within the data path. Some examples are given below:

Instructions number 0 and 1

```
wait until start = 0 and wait until start = 1
```

do not imply any operation (*nop* operation) and the *done* flag is set to 1. The corresponding values of the control signal and of *done* are

start_mult = 0,   *load* = 0,   *shift* = 0,   *en_x_A* = 0, *en_x_B* = 0,   *en_z_A* = 0,   *en_z_B* = 0,   *en_R* = 0, *sel_p1* = 00, *sel_p2* = 00, *sel_a1* = 00, *sel_a2* = 00,   *sel_sq* = 00,   *sel_x_A* = 00,   *sel_x_B* = 00, *sel_z_A* = 00, *sel_z_B* = 00, *sel_R* = 0, *done* = 1.

Actually signals *start_mult, load, shift, en_x_A, en_x_B, en_z_A, en_z_B, en_R* must be equal to 0 and *done* equal to 1. The other values do no matter (don't care values) but have been set to 0.

Instruction number 2

```
x_A = 1, z_A = 0, x_B = x_P, z_B = 1, count = 0, done = 0
```

is executed by the data path when *load* = 1, and all register enable inputs are disabled. The *done* flag is set to 0. Thus

start_mult = 0,   *load* = 1,   *shift* = 0,   *en_x_A* = 0, *en_x_B* = 0,   *en_z_A* = 0,   *en_z_B* = 0,   *en_R* = 0, *sel_p1* = 00, *sel_p2* = 00, *sel_a1* = 00, *sel_a2* = 00,   *sel_sq* = 00,   *sel_x_A* = 00,   *sel_x_B* = 00, *sel_z_A* = 00, *sel_z_B* = 00, *sel_R* = 0, *done* = 0.

Instruction number 3

```
if k_{m-i} = 1 then go to 23
```

does not imply any operation (*nop* operation). The *done* flag is set to 0. Thus

start_mult = 0,   *load* = 0,   *shift* = 0,   *en_x_A* = 0, *en_x_B* = 0,   *en_z_A* = 0,   *en_z_B* = 0,   *en_R* = 0, *sel_p1* = 00, *sel_p2* = 00, *sel_a1* = 00, *sel_a2* = 00,   *sel_sq* = 00,   *sel_x_A* = 00,   *sel_x_B* = 00, *sel_z_A* = 00, *sel_z_B* = 00, *sel_R* = 0, *done* = 0.

Instruction number 4 includes two operations in parallel;

```
z_B = x_A + z_A
```

is executed when *en_Z_B* = 1,   *sel_a1* = 00, *sel_a2* = 01, *sel_z_B* = 00, and

```
start (Z = x_A·z_B)
```

is executed when *start_mult* = 1, *sel_p1* = 00 and *sel_p2* = 01. The *done flag* is set to 0. Thus

start_mult = 1,   *load* = 0,   *shift* = 0,   *en_x_A* = 0, *en_x_B* = 0,   *en_z_A* = 0,   *en_z_B* = 1,   *en_R* = 0, *sel_p1* = 00, *sel_p2* = 01, *sel_a1* = 00, *sel_a2* = 01,   *sel_sq* = 00,   *sel_x_A* = 00,   *sel_x_B* = 00, *sel_z_A* = 00, *sel_z_B* = 00, *sel_R* = 0, *done* = 0.

A complete VHDL model *scalar_product.vhd* is available at the Authors' web site.

2.5.3   Test

Some information about finite binary fields and elliptic curves are given in Appendices A and B. In this example, the binary field is $GF(2^{163})$ and consists of all binary polynomials of degree smaller than 163, with operations modulo a polynomial $f(z)$ of degree $m = 163$:

$$f(z) = z^{163} + z^7 + z^6 + z^3 + 1.$$

Then, a particular elliptic curve EC is defined as follows: it consists of all pairs $(x, y) \in GF(2^{163}) \times GF(2^{163})$ of binary polynomials such that $y^2 + xy = x^3 + x^2 + 1$, plus a particular element ∞ called *element at infinity*:

$$EC = \{(x,y) \in GF(2^{163}) \times GF(2^{163}) | y^2 + xy = x^3 + x^2 + 1\} \cup \{\infty\}.$$

An addition operation can be defined so that EC is a commutative group (Appendix B), being ∞ the neutral element. Then, given a point P of EC and a natural k, the scalar product kP is defined by

$$kP = P + P + \cdots + P(k\ times),$$
$$\forall\, k > 0 \text{ and } 0\,P = \infty.$$

A binary polynomial of degree <163 can be represented by its 163 binary coefficients. Thus, the elements of $GF(2^{163})$ as well as the scalar values k are 163-bit vectors that can also be represented as 41-digit hexadecimal vectors.

To check the working of the circuit, the following values of $P = (x_P, y_P)$ and k are used:

$x_P = $ 2fe13c0537bbc11acaa07d793de4e6d5e5c94eee8,
$y_P = $ 289070fb05d38ff58321f2e800536d538ccdaa3d9,
$k = $ 40000000000000000000020108a2e0cc0d99f8a5ef.

Actually, in Algorithm 2.1 the value of y_P is only used by the final (and not implemented) *last_step* procedure. So, its value is not necessary to simulate the circuit of Sect. 2.5.2.

The simulation of the circuit with the previously defined values of x_P and k gives the following results:

$x_A = $ 1d538b8105663e13c972bf682b49975f7a5fd6345
$z_A = $ 4ae93681fa9e59e7a7aa2b2592ba6e92dcb7d4674
$x_B = $ 2758e50c38d039b358daf65e05bdd89f8fb1e4a1a
$z_B = $ 0000000000000000000000000000000000000000

Algorithm 2.4 is an iteration that is executed m times, and each iteration step includes five multiplications. The interleaved multiplication algorithm is also an iteration that is executed m times. Thus, a lower bound of the number of cycles is $5\,m^2$, in this case $5 \cdot 163^2 = 132{,}845$ cycles.

A test bench *test_scalar_product.vhd* is available at the Authors' web site. The final cycles of the test bench simulation results are shown in Fig. 2.25.

2.6 Comments

All along the two first chapters important concepts have been presented. The starting point of this study of digital circuits is the observation that a practical, rigorous, unambiguous specification method is an algorithm defined in some language —pseudocode and VHDL in the proposed examples and exercises. In Chap. 1, the classical partition of the system into data path and control unit has been presented. In Chap. 2, part of the solution space within which the system designer must move in order to get good circuits is explored. The designer must deduce from the algorithms to be implemented which are the necessary computation resources; they could be standard components, virtual IP components, completely new components that must be developed "from scratch," and others. A first decision that has a direct impact on the cost and performance of the resulting circuit is the scheduling of the operations. From the chosen schedule, not only the computation time of the circuit is deduced, but also the number of resources of each type, the number and type of memory elements, the number and type of connection resources.

A rather long example (Algorithm 2.1 without the final procedure) has been studied. Virtual components (in this case, VHDL models available in a free access Web site) have been used. Several schedules have been proposed with different trade-offs between cost (basically the number of multipliers) and the computation time. The schedule of Fig. 2.12 has been chosen. It permits to implement the algorithm with only one multiplier. The final result is a synthesizable VHDL model of the circuit. However, other circuits could have been developed. Two of them are proposed as Exercises 2 and 3 (Sec. 2.7).

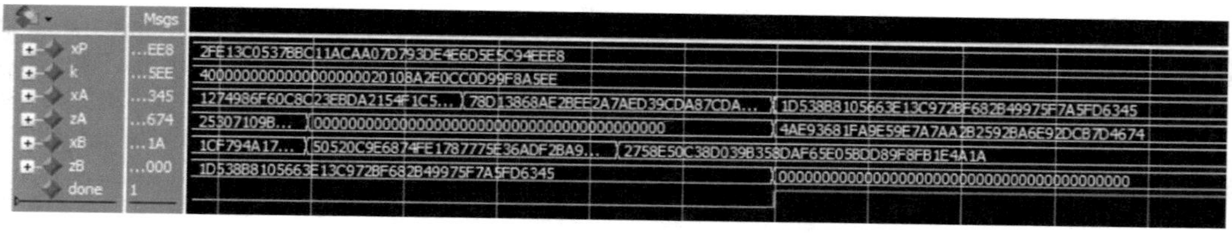

Fig. 2.25 Algorithm 2.4 implementation (ModelSim Starter Edition, courtesy of Mentor Graphics)

Obviously, operation scheduling is fundamental and determines many characteristics of the resulting circuit. However, the solution space exploration is not limited to the finding of the best schedule(s) with respect to some initially specified characteristics. Additional design techniques can be used to improve the circuit performance. An example: in some cases, slight (or not so slight) modifications of the initial algorithm yield significant circuits improvements. In fact, the translation of Algorithm 2.1 to Algorithm 2.2 is an example of such a—in this case trivial—modification, but other example will be seen later. Another example (Fig. 2.18): the multiplexer output register is necessary to permit the parallel execution of two operations ($z_B = x_A + z_A$ and $start(Z = x_A \cdot z_B)$). In the next chapter, it will be seen that the insertion of registers is a technique that sometimes permits to execute operations in parallel and to increase the maximum circuit frequency.

The previous example also suggests some comments in relation to the control unit. The data path is controlled by eighteen signals, nine 1-bit and nine 2-bit signals, in total twenty-seven bits. Obviously, the number of different meaningful commands that the data path can execute is much lower than $2^{27} = 134,217,728$. This observation suggests the use of some types of command encoding that would make easier and clearer the control unit definition.

Another example of possible modification of the circuit of Sect. 2.5: the main circuit (*scalar_product*) consists of a data path and a control unit. One of the data path components (*interleaved_mult*) is also made up of a data path and a control unit, while the *classic_squarer* component is a combinational circuit. An alternative solution is the definition of a data path able to execute all the operations, including those corresponding to the *interleaved_mult* and *classic_squarer* components. The so-obtained circuit could be more efficient than the proposed one as some computation resources could be shared between the three algorithms (field multiplication, squaring and scalar product). Nevertheless, the hierarchical approach consisting of using already existing components is probably safer and allows reducing the development times.

The next chapters are dedicated to several design techniques that permit to improve the circuit performance or to make the design work safer and easier to automatize.

2.7 Exercises

1. Generate several VHDL models of a 7-to-3 counter. Use for that the three options proposed in Sect. 2.1 (Figs. 2.3, 2.4 and 2.5).
2. Implement Algorithm 2.2 using the schedule of Fig. 2.8 so that three multipliers must be used and the computation time is about $2M$ cycles. As before, use the synthesizable source files *classic_squarer.vhd* and *interleaved_mult.vhd* available at the Authors' web site.
3. Implement Algorithm 2.2 using the schedule of Fig. 2.10b so that two multipliers must be used and the computation time is about $3M$ cycles. As before, use the synthesizable source files *classic_squarer.vhd* and *interleaved_mult.vhd* available at the Authors' web site.
4. The following algorithm defines the *last_step* procedure of Algorithm 2.1:

```
if zB = 0 then xA = xP; yA = xP + yP;
else
 xA = xA / zA;
 xB = xB / zB;
 yA = ((xA + xP)[(xA + xP)(xB + xP) + xP² + yP]
 / xP) + yP;
end if;
xR = xA; yR = yR;
```

Use the synthesizable source files *classic_squarer.vhd*, *interleaved_mult.vhd* and *mod_f_binary_division.vhd* (division in $GF(2^m)$) available at the Authors' web site.
5. Design a circuit to compute the greatest common divisor of two natural numbers, based on the following simplified Euclidean algorithm.

```
while a ≠ b loop
  if a > b then a = a - b;
  else b = b - a;
end loop;
gcd = a;
```

6. The distance d between two points (x_1, y_1) and (x_2, y_2) of the (x, y)-plane is equal to $d = ((x_1 - x_2)^2 + (y_1 - y_2)^2)^{0.5}$. Design a circuit that computes d with only one subtractor and one multiplier.

7. Design a circuit that computes the distance between two points (x_1, y_1, z_1) and (x_2, y_2, z_2) of the three-dimensional space.

8. Given a point (x, y, z) of the three-dimensional space, design a circuit that computes the following transformation.

$$\begin{bmatrix} x_t \\ y_t \\ z_t \end{bmatrix} = \begin{bmatrix} a_{11} & a_{21} & a_{31} \\ a_{21} & a_{22} & a_{32} \\ a_{31} & a_{32} & a_{11} \end{bmatrix} \times \begin{bmatrix} x \\ y \\ z \end{bmatrix}$$

9. Design a circuit for computing $z = e^x$ using the formula

$$e^x = 1 + \frac{x}{1!} + \frac{x^2}{2!} + \frac{x^3}{3!} + \cdots$$

10. Design a circuit for computing x^n, where n is a natural, using the following relations: $x^0 = 1$; if n is even then $x^n = (x^{n/2})^2$, and if n is odd then $x^n = x \cdot (x^{(n-1)/2})^2$.

Bibliography

Deschamps JP, Imaña JL, Sutter G (2009) Hardware Implementation of Finite-Field Arithmetic. McGraw-Hill, New York.

Hankerson D, Menezes A, Vanstone S (2004) Guide to Elliptic Curve Cryptography. Springer, New York.

López J, Dahab R (1999) Improved Algorithm for Elliptic Curve Arithmetic in $GF(2^n)$. Lecture Notes in Computer Science 1556:201–212.

A very useful implementation technique, especially for signal processing circuits, is pipelining (De Micheli 1994; Parhami 2000). It consists in inserting additional registers so that the maximum clock frequency and input data throughput are increased. Furthermore, in the case of FPGA implementations, the insertion of pipeline registers has a positive effect on the power consumption.

3.1 Introductory Example

Consider the introductory example of Sect. 2.1. The set of operations (2.5) can be implemented by a combinational circuit (Fig. 2.3) made up of four carry-save adders, with a computation time equal to $3 \cdot T_{FA}$. That means that the minimum clock period of a synchronous circuit including this 7-to-3 counter should be greater than $3 \cdot T_{FA}$, and that the introduction interval between successive data inputs should also be greater than $3 \cdot T_{FA}$. The corresponding circuit is shown in Fig. 3.1a. As previously commented, this is probably a bad circuit because its cost is high and its maximum clock frequency is low.

Consider now the circuit of Fig. 3.1b in which registers have been inserted in such a way that operations scheduled in successive cycles, according to the ASAP schedule of Fig. 2.6a, are separated by a register. The circuit still includes four carry-save adders, but the minimum clock

period of a synchronous circuit including this counter must be greater than T_{FA}, plus the setup and hold times of the registers, instead of $3 \cdot T_{FA}$. Furthermore, the minimum data introduction interval is now equal to T_{clk}: as soon as a_1, a_2, b_1 and b_2 have been computed, their values are stored within the corresponding output register, and a new computation, with other input data, can start; at the same time, new computations of c_1 and c_2, and of d_1 and d_2 can also start. Thus, at time t, three operations are executed in parallel:

$$(a_1(t), a_2(t)) = CSA(x_1(t), x_2(t), x_3(t)),$$
$$(b_1(t), b_2(t)) = CSA(x_4(t), x_5(t), x_6(t));$$
$$(c_1(t), c_2(t)) = CSA(b_1(t-1),$$
$$b_2(t-1), x_7(t-1));$$
$$(y_1(t), y_2(t)) = CSA(a_1(t-2), a_2(t-2),$$
$$c_1(t-1)), y_3(t)$$
$$= c_2(t-1);$$

so that

$$y_1(t) + y_2(t) + y_3(t)$$
$$= a_1(t-2) + a_2(t-2) + c_1(t-1)$$
$$+ c_2(t-1)$$
$$= x_1(t-2) + x_2(t-2) + x_3(t-2)$$
$$+ b_1(t-2) + b_2(t-2) + x_7(t-2)$$
$$= x_1(t-2) + x_2(t-2) + x_3(t-2)$$
$$+ x_4(t-2) + x_5(t-2)$$
$$+ x_6(t-2) + x_7(t-2).$$

© Springer Nature Switzerland AG 2019
J.-P. Deschamps et al., *Complex Digital Circuits*,
https://doi.org/10.1007/978-3-030-12653-7_3

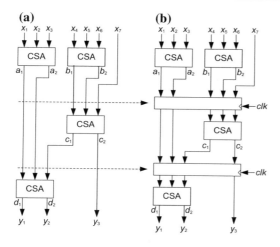

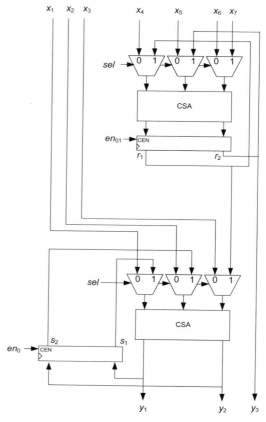

Fig. 3.1 **a** Combinational circuit. **b** Pipelined circuit

The total computation time is equal to $2 \cdot T_{clk} + T_{FA} \approx 3 \cdot T_{clk}$, and a new set of input data can be inputted every cycle.

To summarize, assuming that the setup and hold times are negligible,

$$T_{clk} > T_{FA}, latency = 3 \cdot T_{clk}, r = T_{clk},$$

where *latency* is the total computation time and r is the minimum data introduction interval (Definition 3.1).

Another implementation with two carry-save adders has been shown in Fig. 2.4. This same circuit, with a different drawing, is also shown in Fig. 3.2. It works in three steps:

```
0 : (s₁, s₂)= CSA(x₁, x₂, x₃),(r₁, r₂)= CSA(x₄, x₅, x₆);
1 : (r₁, r₂)= CSA(r₁, r₂, x₇);
2 : (y₁, y₂)= CSA(s₁, s₂, r₁), y₃= r₂;
```

and is based on the ASAP schedule of Fig. 2.6a. The minimum clock period of a synchronous circuit including this counter must be greater than T_{FA}, plus the setup and hold times of the registers. The total computation time is equal to $3 \cdot T_{clk}$, and the minimum data introduction interval is also equal to $3 \cdot T_{clk}$. A VHDL model *seven_-to_three_seq_reg.vhd* is available at the Authors' web site. In this model, an output register, enabled during the third step, has been added in order to synchronize the three output signals. Part

Fig. 3.2 Three-cycle implementation of a 7-to-3 counter

of the simulation result is shown in Fig. 3.3: the circuit computes

$$17 + 18 + 19 + 16 + 17 + 18 + 19$$
$$= 22 + 64 + 38,$$
$$33 + 34 + 35 + 32 + 33 + 34 + 35$$
$$= 38 + 128 + 70,$$

in three cycles and starts a new computation every three cycles.

Consider now the circuit of Fig. 3.4. It consists of a modification of the circuit of Fig. 3.2 in which an additional pipeline register has been inserted. This new circuit is made up of two stages separated by the pipeline register. Within every stage, the operations are executed in two cycles. The first stage executes the two following successive steps:

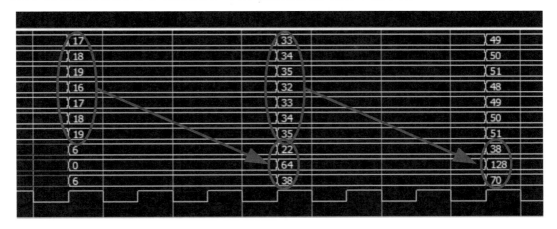

Fig. 3.3 Simulation of a 7-to-3 counter (courtesy of Mentor Graphics)

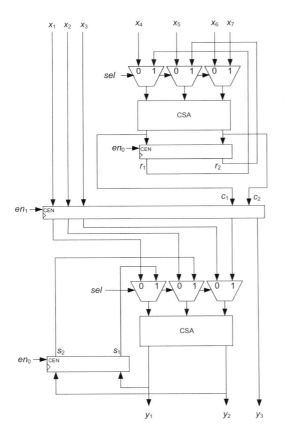

Fig. 3.4 Two-stage two-cycle implementation

$0 : (r_1(t), r_2(t)) = \mathrm{CSA}(x_4(t), x_5(t), x_6(t));$

$1 : (c_1(t), c_2(t)) = \mathrm{CSA}(r_1(t), r_2(t), x_7(t));$

and the second stage executes

$0 : (s_1(t), s_2(t)) = \mathrm{CSA}(x_1(t-1), x_2(t-1), x_3(t-1));$

$1 : (y_1(t), y_2(t)) = \mathrm{CSA}(s_1(t), s_2(t), c_1(t-1)),$
$$y_3(t) = c_2(t-1);$$

At the end of the second step,

- The first stage generates $r_1(t)$ and $r_2(t)$ such that

$$c_1(t) + c_2(t) = x_4(t) + x_5(t) + x_6(t) + x_7(t);$$

- The second stage generates $y_1(t)$, $y_2(t)$ and $y_3(t)$ such that

$$\begin{aligned}
y_1(t) &+ y_2(t) + y_3(t) \\
&= x_1(t-1) + x_2(t-1) + x_3(t-1) \\
&\quad + c_1(t-1) + c_2(t-1) \\
&= x_1(t-1) + x_2(t-1) + x_3(t-1) \\
&\quad + x_4(t-1) + x_5(t-1) \\
&\quad + x_6(t-1) + x_7(t-1).
\end{aligned}$$

An output register, enabled during the second cycle, synchronizes the three output signals.

The minimum clock period of a synchronous circuit including this counter must be greater than T_{FA}, plus the setup and hold times of the registers. The total computation time is equal to $4 \cdot T_{clk}$ (two stages and two cycles per stage), but the two stages work in parallel, the first with input data $x_4(t)$ to $x_7(t)$ and the second with input data $x_1(t-1)$ to $x_3(t-1)$, so that the minimum data introduction interval is now equal to $2 \cdot T_{clk}$. A VHDL model *seven_to_three_pipe.vhd* is available at the Authors' web site. Part of the simulation result is shown in Fig. 3.5: the circuit computes

$$17 + 18 + 19 + 16 + 17 + 18 + 19$$
$$= 22 + 64 + 38,$$
$$33 + 34 + 35 + 32 + 33 + 34 + 35$$
$$= 38 + 128 + 70,$$

in four cycles and starts a new computation every two cycles.

The circuit of Fig. 3.4 includes two carry-save adders, and its timing constraints are the following:

$$T_{clk} > T_{FA} + T_{multiplexor}, latency = 4 \cdot T_{clk}, r = 2 \cdot T_{clk}$$

where *latency* is the total computation time and r is the minimum data introduction interval.

To summarize, the sequential circuit of Fig. 3.2, whose simulation result is shown in Fig. 3.3, computes the 7-to-3 counter function in three cycles and can start a new computation every three cycles, while the pipelined circuit of Fig. 3.4, whose simulation result is shown in Fig. 3.5, computes the same function in four cycles but can start a new computation every two cycles.

Definition 3.1 The main parameters of a pipelined circuit are the following.

- The *latency* (also called *delay* or *response time*) is the total delay between the introduction of a new set of input data and the generation of the corresponding output results. If the circuit consists of n pipeline stages and all stages are executed in s clock period T_{clk}, then the latency is equal to $n \cdot s \cdot T_{clk}$. Thus, the latency of the circuit of Fig. 3.1b ($n = 3$, $s = 1$) is equal to $3 \cdot T_{clk}$, the latency of the circuit of Fig. 3.2 (actually a non-pipelined circuit so that $n = 1$, $s = 3$) is equal to $3 \cdot T_{clk}$ and the latency of the circuit of Fig. 3.4 ($n = 2$, $s = 2$) is equal to $4 \cdot T_{clk}$.
- The *pipeline rate* (also called *pipeline period*) is the data introduction interval. If all stages are executed in s clock cycles, the pipeline rate is equal to $s \cdot T_{clk}$. Thus, the pipeline rate of the circuit of Fig. 3.1.b is equal to T_{clk}, the pipeline rate of the circuit of Fig. 3.2 is equal to $3 \cdot T_{clk}$ and the pipeline rate of the circuit of Fig. 3.4 is equal to $2 \cdot T_{clk}$.

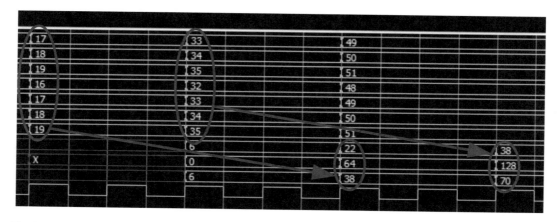

Fig. 3.5 Simulation of the pipelined circuit (courtesy of Mentor Graphics)

- The *throughput* (also called *speed*, *bandwidth* or *production*) is the number of input data processed per time unit. Consider a sequence of m successive input data. The computation time of the first output data is equal to the latency, that is, $n \cdot s \cdot T_{clk}$. Then, a new output data is generated every $r = s \cdot T_{clk}$ time units. Thus, the total computation time is equal to $n \cdot s \cdot T_{clk} + (m - 1) \cdot r$ and the throughput is equal to $m/(n \cdot s \cdot T_{clk} + (m - 1) \cdot r)$. So, for great numbers of processed data ($m \to \infty$) the throughput is equal to the inverse $1/r$ of the pipeline rate r.

The main positive result of pipelining is an important increase of the throughput (or bandwidth) without a great increase of the cost. To understand that point, consider the generic circuit of Fig. 3.6a. It consists of n subcircuits whose latencies are approximately equal, say T time units. They could be combinational circuits or sequential circuits that compute their function in s clock periods in which case $T = s \cdot T_{clk}$. The total computation time is equal to $n \cdot T$, and the throughput (number of input data processed per time unit) is equal to $1/n \cdot T$. The circuit of Fig. 3.6b is the pipelined version of the preceding. Taking into account the setup times and hold times of the pipeline registers, the period of the pipeline register clock must be a bit longer than T, say $T + \delta$ time units. The total processing time of a sequence of m input data is equal to $n \cdot (T + \delta) + (m - 1) \cdot (T + \delta)$ time units, so that the throughput is equal to $m/(n \cdot (T + \delta) + (m - 1) \cdot (T + \delta))$. For great m ($m \to \infty$), the throughput is equal to $1/(T + \delta)$. The *speedup factor* that is the relation between the throughput of the pipelined circuit (Fig. 3.6b) and the throughput of the initial circuit (Fig. 3.6a) is equal to

$$speedup\,factor = n \cdot T/(T + \delta)$$
$$= n/(1 + \alpha)\ where\ \alpha = \delta/T.$$
$$(3.1)$$

Generally, $\alpha \ll 1$ as it is the relation between the sum of the setup and hold times of a register and the computation time of a subcircuit of the original circuit (Fig. 3.6a).

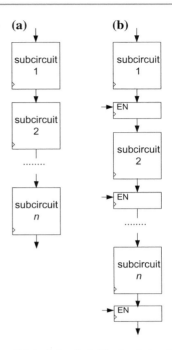

Fig. 3.6 a Original circuit. **b** Pipelined circuit

Thus, the throughput of the circuit has been multiplied by $n/(1 + \alpha) \cong n$, that is, the number of pipeline stages. On the other hand, the additional cost of the pipelined circuit corresponds to the pipeline registers. Assume that the n subcircuits of Fig. 3.6a have approximately the same cost C and that the pipeline registers are identical (same number of bits). Then, the cost without pipelining is equal to $n \cdot C$, while the cost of the pipelined circuit is equal to $n \cdot (C + C_{reg})$. The cost increase is equal to

$$(C + C_{reg})/C = 1 + \beta\ where\ \beta = C_{reg}/C.$$
$$(3.2)$$

If all n subcircuits are relatively complex circuits, so that $C \gg C_{reg}$, then $\beta \ll 1$ and the cost increase is low.

It is worthwhile to comment that pipelining is a very effective technique in the case of an FPGA implementation of the circuit.

- The basic cell of a field-programmable gate arrays includes a flip-flop, so that the insertion of pipeline registers does not necessarily increase the total cost, computed in terms of used basic cells. The pipeline registers could consist of flip-flops not used in the non-pipelined version.

- Most FPGA families also permit to implement with lookup tables (LUTs) those registers that do not need reset signals. This can be another cost-effective option.
- The insertion of pipeline registers also has a positive effect on the power consumption: The presence of synchronization barriers all along the circuit drastically reduces the number of generated spikes.

3.2 Segmentation

In the preceding examples of pipelined circuits, the partition of the circuit into several subcircuits (Figs. 3.1, 3.2 and 3.6) was assumed to have been defined in advance. The practical problem that the circuit designer is faced with is the following: given an initial non-pipelined circuit, how can it be partitioned in such a way that an efficient pipelined implementation could be considered.

According to the general design strategy proposed in this course, it is assumed that the initial circuit has been developed starting from an algorithm and from the study of the precedence relations between the algorithm operations. Then, given a computation scheme and its precedence graph G, a *segmentation* of G is an ordered partition $\{S_1, S_2, ..., S_k\}$ of G. The segmentation is *admissible* if it respects the precedence relation. This means that if there is an arc from $op_J \in S_i$ to op_M then either op_M belongs to the same segment S_i or it belongs to a different segment S_j with $j > i$.

In fact, the elements S_i of G will correspond to pipeline stages and the order of the elements of G will correspond to the order of the corresponding stages. Thus, to be admissible the chosen segmentation must respect the following rule: if an operation is executed by some pipeline stage, say number i, then all the data it generates could only be used within this same stage or within stage number j with $j > i$.

Two examples are shown in Fig. 3.7 in which the segments are separated by dotted lines.

The segmentation of Fig. 3.7a, that is $G_1 = \{op_1, op_2\}$, $G_2 = \{op_3, op_4\}$, $G_3 = \{op_5, op_6\}$, is admissible, while that of Fig. 3.7b, that is $G_1 = \{op_1, op_3\}$, $G_2 = \{op_2, op_5\}$, $G_3 = \{op_4, op_6\}$, is not (there is an arc $op_2 \rightarrow op_3$ from G_2 to G_1).

Once an admissible partition has been defined, every segment can be synthesized separately, using the same methods as before (scheduling, resource assignment). In order to assemble the complete circuit, additional registers are inserted: if an arc of the precedence graph crosses the line that separates segments i and $i + 1$, then a register must be inserted; it will store the output data generated by segment i that in turn are input data to segment $i + 1$. As an example, the structure of the circuit corresponding to Fig. 3.7a is shown in Fig. 3.8.

Assume that C_i and T_i are the cost and computation time of segment i. The cost of the complete circuit is equal to

$$C = C_1 + C_2 + \ldots + C_k + C_{registers} \qquad (3.3)$$

where $C_{registers}$ represents the total cost of the pipeline registers.

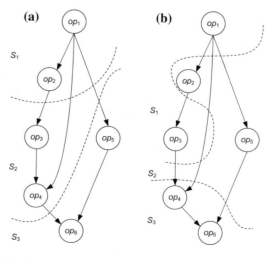

Fig. 3.7 a Admissible segmentation. **b** Non-admissible segmentation

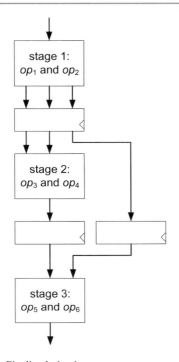

Fig. 3.8 Pipelined circuit

The pipeline registers are synchronized by a clock signal whose period $T_{pipeline}$ must be greater than all segment computation times:

$$T_{pipeline} \geq max\{T_1, T_2, \cdots, T_k\} + T_{registers} \quad (3.4)$$

where $T_{registers}$ is the additional delay introduced by the pipeline registers (setup and hold times).

The latency of the complete circuit is equal to

$$T = k \cdot T_{pipeline} \geq k \\ \cdot [max\{T_1, T_2, \cdots, T_k\} + T_{registers}] \quad (3.5)$$

and the pipeline rate time δ, that is the time interval between successive data inputs, is

$$\delta = T_{pipeline} \geq max\{T_1, T_2, \cdots, T_k\} + T_{registers} \\ (3.6)$$

Example 3.1 Consider a three-dimensional space defined as follows:

$$S = \{0, 1, 2, \cdots, 2^M - 1\}^3 \quad (3.7)$$

for some natural M. The space points are represented under the form (x, y, z) where x, y and z are M-bit naturals. The circuit to be designed computes the distance between two points (x_1, y_1, z_1) and (x_2, y_2, z_2):

$$distance = \Big[|x_2 - x_1|^2 + |y_2 - y_1|^2 \\ + |z_2 - z_1|^2\Big]^{1/2} \quad (3.8)$$

The maximum value of *distance* is

$$[3 \cdot (2^M - 1)^2]^{1/2} = 3^{1/2} \cdot (2^M - 1) < 2^{M+1} \\ (3.9)$$

so that it is an $(M + 1)$-bit natural.

According to (3.7), the center of S is point $(2^{M-1}, 2^{M-1}, 2^{M-1})$ and the minimum distance between two points is 1. If the circuit under development were included within a system that process points belonging to a three-dimensional space whose center is $(0, 0, 0)$ and whose point coordinates are signed fixed-point numbers, previous translation and scaling operations should have been previously executed.

The following algorithm computes (3.8):

```
a = | x₂ - x₁ | ;
b = | y₂ - y₁ | ;
c = | z₂ - z₁ | ;
d = a² ;
e = b² ;
f = c² ;
g = d + e + f ;
distance = g^(1/2) ;
```

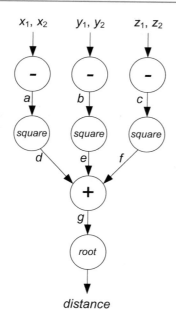

Fig. 3.9 Distance between two points

The corresponding precedence graph is shown in Fig. 3.9.

Several admissible segmentations are shown in Fig. 3.10.

Before choosing a pipeline configuration, some previous decisions must be taken. Two predefined components will be used: *shift_and_add_multiplier.vhd* and *SquareRoot.vhd*. Complete and synthesizable source files are available at the Authors' web site. They are considered as predefined IP components (intellectual property components) whose main characteristics are the following.

- The *shift_and_add_multiplier* component computes $z = x \cdot y + u + v$, where x and u are n-bit naturals, y and v are m-bit naturals and z is an $(n + m)$-bit natural. The computation is executed in m cycles with a minimum clock period approximately equal to the delay of an n-bit adder. In this circuit, inputs u and v will be connected to constant values 0.
- The data input of the *SquareRoot* component is a $2n$-bit natural x. Its outputs are an n-bit natural *root* and an $(n + 1)$-bit natural *remainder* such that $x = root^2 + remainder$ where $remainder \leq 2 \cdot root$. The computation is executed in n cycles, and the minimum clock period is approximately equal to the delay of an n-bit adder.

Observe that if $x = root^2 + remainder$ and $remainder \leq 2 \cdot root$, then

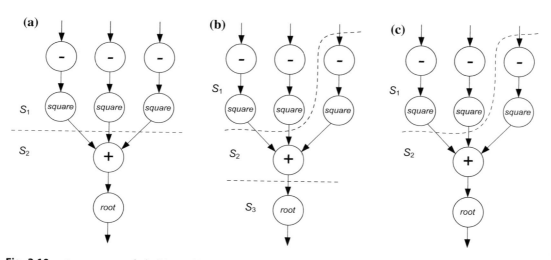

Fig. 3.10 **a** two-segment admissible partition. **b** three-segment admissible partition. **c** two-segment admissible partition

$$root^2 = x - remainder,$$
$$(root + 1)^2 = root^2 + 2 \cdot root + 1$$
$$= (x - remainder) + 2 \cdot root$$
$$+ 1 \geq x + 1,$$

so that (*remainder* is natural)

$$root^2 \leq x \leq (root + 1)^2 - 1 < (root + 1)^2. \quad (3.10)$$

According to (3.10), $root = \lfloor x^{1/2} \rfloor$, that is, the integer square root of x.

Thus, *shift_and_add_multiplier* components are used to compute

$$d = a \cdot a + 0 + 0 = a^2, e = b \cdot b + 0 + 0$$
$$= b^2 \text{ and } f = c \cdot c + 0 + 0 = c^2$$

(Fig. 3.9), where a, b and c are M-bit naturals, and d, e and f are $2M$-bit naturals. The value of parameters m and n is $m = n = M$, so that $m + n = 2M$, and the computation is executed in M cycles with a minimum clock period approximately equal to the delay of an M-bit adder.

A *SquareRoot* component is used to compute

$$distance = g^{1/2}$$

(Fig. 3.9), where *distance* is an $(M + 1)$-bit natural (Eq. 3.9) so that g is a $(2M + 2)$-bit natural and the value of parameter n is $M + 1$. The computation is executed in $n = M + 1$ cycles with a minimum clock period approximately equal to the delay of an $(M + 1)$-bit adder.

The differences $a = |x_2 - x_1|$, $b = |y_2 - y_1|$, $c = |z_2 - z_1|$ where x_1, x_2, y_1, y_2, z_1 and z_2 are M-bit naturals will be computed by combinational circuits whose delays are approximately equal to the delay of an M-bit adder.

The sum $g = d + e + f$ (Fig. 3.9) where d, e and f are $2M$-bit naturals will also be computed by a combinational circuit whose delay is approximately equal to the delay of a $2M$-bit adder.

To simplify the circuit design, it is assumed that

- The clock period T_{clk} is greater than the delay of a $2M$-bit adder.
- The computation times of the *shift_and_add* and *SquareRoot* components are equal to $N \cdot T_{clk}$ where the number N of cycles is (slightly) greater than $M + 1$ (so as to take into account initial *start* and final *done* cycles).

To summarize, with the preceding assumptions,

- a, b, c and g are computed in 1 cycle.
- d, e, f and *distance* are computed in N cycles.

Consider the segmentation of Fig. 3.10a to which corresponds a two-stage pipeline. Within each stage, the schedule of the operations must be defined. In the case of Fig. 3.11a, the first stage is scheduled as follows: the three differences a, b and c are computed during cycle 1, and the three squares $d = a^2$, $e = b^2$ and $f = c^2$ are computed during cycles 2 to $N + 1$. During cycle $N + 2$, the stage 1 results are transferred to stage 2. The second stage is scheduled as follows: the sum $g = d + e + f$ is computed during cycle 1, and the square root *distance* $= g^{1/2}$ is computed during cycles 2 to $N + 1$. During cycle $N + 2$, the stage 2 results are transferred to the circuit output.

This first solution is a pipeline with two stages, each of them constituted of $N + 2$ cycles. Thus,

$$latency = 2 \cdot (N + 2) \cdot T_{clk} 0 \cong 2 \cdot N \cdot T_{clk}, \quad (3.11a)$$

$$pipeline\ rate = (N + 2) \cdot T_{clk} \cong N \cdot T_{clk}, \quad (3.11b)$$

$$cost > 3 \cdot C_{square} + C_{root}. \quad (3.11c)$$

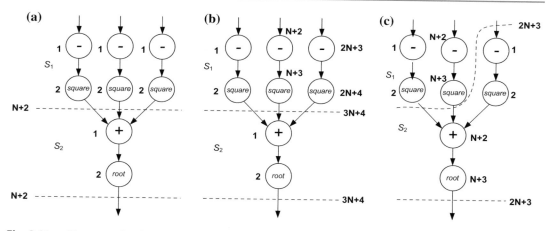

Fig. 3.11 **a** First example of schedule. **b** Second example of schedule. **c** Third example of schedule

In (3.11c), only the cost of the more complex components has been taken into account. The actual cost also includes the register, adder, connection and control unit costs.

In the case of Fig. 3.11b, the first stage is scheduled as follows: a is computed during cycle 1, the square $d = a^2$ is computed during cycles 2 to $N + 1$, b is computed during cycle $N + 2$, the square $e = b^2$ is computed during cycles $N + 3$ to $2N + 2$, c is computed during cycle $2N + 3$, the square $f = c^2$ is computed during cycles $2N + 4$ to $3N + 3$, and the stage 1 results are transferred to stage 2 during cycle $3N + 4$. The second stage is scheduled as follows: the sum $g = d + e + f$ is computed during cycle 1, and the square root $distance = g^{1/2}$ is computed during cycles 2 to $N + 1$. The stage 2 results are available at the beginning of cycle $N + 2$. However, all pipeline registers must be updated at the same time in order to maintain the data flow synchronization, so that the stage 2 results are transferred to the circuit output during cycle number $3N + 4$.

This second solution is a pipeline with two stages and $s = 3N + 4$ cycles per stage. Thus,

$$latency = 2 \cdot (3N + 4) \cdot T_{clk} \cong 6 \cdot N \cdot T_{clk},$$
$$(3.12a)$$

$$pipeline\ rate = (3N + 4) \cdot T_{clk}, \cong 3 \cdot N \cdot T_{clk},$$
$$(3.12b)$$

$$cost > C_{square} + C_{root}.$$
$$(3.12c)$$

Comparing with the precedent solution (Eq. 3.11a), the cost has been lowered (two complex components instead of four), but the pipeline rate has been multiplied by three (the throughput has been divided by three).

Consider now the segmentation of Fig. 3.10c, to which corresponds a two-stage pipeline, with the stage schedules of Fig. 3.11c. The first stage is scheduled as follows: a is computed during cycle 1, the square $d = a^2$ is computed during cycles 2 to $N + 1$, b is computed during cycle $N + 2$, the square $e = b^2$ is computed during cycles $N + 3$ to $2N + 2$, and the stage 1 results are transferred to stage 2 during cycle $2N + 3$. The second stage is scheduled as follows: c is computed during cycle 1, the square $f = c^2$ is computed during cycles 2 to $N + 1$, the sum $g = d + e + f$ is computed during cycle $N + 2$, and the square root $distance = g^{1/2}$ is computed during cycles $N + 3$ to $2N + 2$. The stage 2 results are transferred to the circuit output during cycle number $2N + 3$.

This third solution is a pipeline with two stages and $2N + 3$ cycles per stage. Thus,

$$latency = 2 \cdot (2N+3) \cdot T_{clk} \cong 4 \cdot N \cdot T_{clk}, \tag{3.13a}$$

$$pipeline\ rate = (2N+3) \cdot T_{clk}, \cong 2 \cdot N \cdot T_{clk}, \tag{3.13b}$$

$$cost > 2 \cdot C_{square} + C_{root}. \tag{3.13c}$$

Comparing with the precedent solutions (Eqs. 3.11a and 3.12a), the cost (three complex components) and the pipeline rate have intermediate values.

It is also interesting to evaluate the circuit performance of a one-stage circuit (actually a non-pipelined circuit). Consider the ASAP schedule of Fig. 3.12.

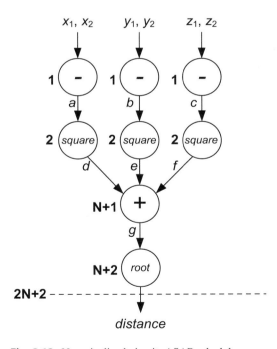

Fig. 3.12 Non-pipelined circuit, ASAP schedule

$$latency = (2N+2) \cdot T_{clk} \cong 2 \cdot N \cdot T_{clk}, \tag{3.14a}$$

$$pipeline\ rate = latency \cong 2 \cdot N \cdot T_{clk}, \tag{3.14b}$$

$$cost > 3 \cdot C_{square} + C_{root}. \tag{3.14c}$$

Table 3.1 is a comparison of the preceding options.

Obviously, there are other solutions. For example, the admissible partition of Fig. 3.10b could be considered. However, it is probably not a good option. If the first stage is executed with two multipliers, then the number of cycles per stage is approximately equal to N but four complex components must be used. Thus, the rate is approximately equal to $N \cdot T_{clk}$, the cost is greater than $4 \cdot C_{complex-component}$, but the latency is equal to $3 \cdot rate \cong 3 \cdot N \cdot T_{clk}$, a longer delay than in the second row of Table 3.1. If the first stage is executed with one multiplier, then the number of cycles to execute the first stage is approximately equal to $2N$. The rate is approximately equal to $2 \cdot N \cdot T_{clk}$, the cost is greater than $3 \cdot C_{complex-component}$, but the latency is approximately equal to $3 \cdot rate \cong 6 \cdot N \cdot T_{clk}$, a longer delay than in the fourth row of Table 3.1.

To conclude this example, the circuit corresponding to the schedule of Fig. 3.11c is implemented. A block diagram of the data path is shown in Fig. 3.13. An additional control unit generates the control signals *sel*, *start*1, *en*, *en_pipe*, *start*21 and *start*22. It is a mod $2N + 3$ counter plus a decoder (Table 3.2) that generates the control signals in function of the counter state and according to the schedule of Fig. 3.11c.

A VHDL model *distance.vhd* is available at the Authors' web site. Part of the simulation result with $M = 8$ and $N = 12$ is shown in Fig. 3.14. The circuit computes the distance between points (11, 10, 9) and (7, 38, 5), points (0, 0, 0)

Table 3.1 Comparison of pipelined circuits

	Latency	Rate	Cost
One stage, Fig. 3.12	$2 \cdot N \cdot T_{clk}$	$2 \cdot N \cdot T_{clk}$	$4 \cdot C_{complex-component}$
Two stages, Fig. 3.11a	$2 \cdot N \cdot T_{clk}$	$N \cdot T_{clk}$	$4 \cdot C_{complex-component}$
Two stages, Fig. 3.11b	$6 \cdot N \cdot T_{clk}$	$3 \cdot N \cdot T_{clk}$	$2 \cdot C_{complex-component}$
Two stages, Fig. 3.11c	$4 \cdot N \cdot T_{clk}$	$2 \cdot N \cdot T_{clk}$	$3 \cdot C_{complex-component}$

Fig. 3.13 Distance computation

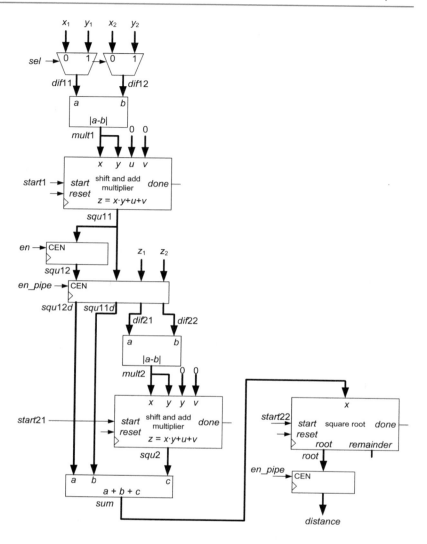

Table 3.2 Command decoder

State n°	sel	start_1	en	en_pipe	start21	start22
0	0	0	0	0	0	0
1	0	1	0	0	1	0
2 to $N-1$	0	0	0	0	0	0
N	0	0	1	0	0	0
N + 1	1	0	0	0	0	0
N + 2	1	0	0	0	0	1
N + 3 to 2 N + 1	1	0	0	0	0	0
2N + 2	0	0	0	1	0	0

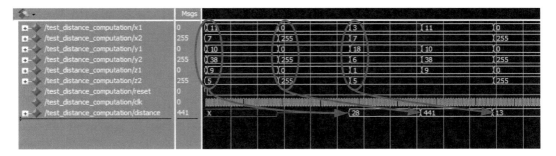

Fig. 3.14 Simulation of the distance computation (courtesy of Mentor Graphics)

and (255, 255, 255), and points (3, 18, 1) and (7, 6, 5):

$$\lfloor[|11-7|^2+|10-38|^2+|9-5|^2]^{1/2}\rfloor = 28,$$
$$\lfloor[|0-255|^2+|0-255|^2+|0-255|^2]^{1/2}\rfloor = 441,$$
$$\lfloor[|3-7|^2+|18-6|^2+|1-5|^2]^{1/2}\rfloor = 13.$$

The latency is equal to $4N + 6 = 54$ cycles, and the pipeline rate is equal to $2N + 3 = 27$ cycles.

Comment 3.1

- Pipelining has been presented in the case of computation schemes (algorithm without branches and loops). How can the case of more complex algorithms be treated, for instance algorithms including loops? In Chap. 2, an elliptic curve cryptography algorithm has been partially implemented. It consists of a loop whose body includes a branch (Algorithm 2.1). The proposed method consisted in implementing a data path able to execute both branches of the loop body and to define a control unit that executes the complete algorithm using this data path as a specific processor. This data path (Fig. 2.23) can be segmented into several pipeline stages. An example, with five stages, is given in Deschamps et al. (2012, Example 3.1). Nevertheless, the initial data of every loop body execution is the results of the preceding loop body execution. In consequence, new input data is not available until the preceding loop body execution is completed. Thus, the pipeline rate is equal to the latency. Two advantages of the pipelined circuit are that the clock frequency has been incremented (multiplied by the number of pipeline stages) and that the power consumption might have been reduced because the synchronization barriers (pipeline registers) reduce the generation of spikes, but the throughput is the same as that of the non-pipelined circuit. To increase the throughput, the loop must be (at least partially) flatten. This technique will be studied in Chap. 4.

- More flexible pipeline circuits (elastic pipelines) can also be considered: instead of connecting the stages by means of pipeline registers, all of them controlled by the same clock signal, a more flexible configuration uses first-in first-out (FIFO) memories instead of registers. Then, the data flow control relies on the following rule: a pipeline stage can process data if its input FIFO is not empty, and its output FIFO is not full. This technique will be studied in Chap. 5.

3.3 Circuit Transformations

Often, pipelining techniques are used to improve the working of an already existing circuit. In such a case, the pipelined circuit is not developed "from scratch" and the designer work consists of a rather simple modification of the previously developed circuit. Two cases are considered in this section: combinational circuits and digital signal processing circuits.

3.3.1 Combinational to Pipelined Transformations

Consider a combinational circuit made up of relatively small blocks, all of them with nearly equal delays. Assume that the longest input-to-output path (the critical path) goes through n blocks. Then, the total computation time of this combinational circuit is equal to $n \cdot T$ time units. If this circuit is used as a computation resource of a synchronous circuit, then the clock cycle must be greater than $n \cdot T$, and in some cases it could be a too long time (a too low frequency). In order to increase the clock frequency, as well as to reduce the minimum time interval between successive data inputs, the solution is pipelining. As the combinational version already exists, it is no longer necessary to use the general method of Sect. 3.2 and the combinational circuit can be directly segmented into stages. Actually, an example has already been shown (Fig. 3.1). Consider another generic example.

Example 3.2 The iterative circuit of Fig. 3.15a is made up of twelve identical blocks (cells), each of them with a maximum delay of t_{cell} seconds. Assume that it is part of a synchronous circuit, and that all its inputs come from register outputs and all its outputs go to register inputs. The registers have minimum setup and propagation times equal to t_{SU} and t_P time units, respectively, and all connections are assumed to have the same propagation delay $t_{connection}$. Then, the longest paths between the input register and the output register include six cells and seven connections.

One of the critical paths is shown in Fig. 3.15b. Thus, the minimum clock cycle T_{CLK} must satisfy the following relation:

$$T_{CLK} > 6 \cdot t_{cell} + 7 \cdot t_{connection} + t_{SU} + t_P.$$

$$(3.15)$$

If the period defined by condition (3.15) is too long, the circuit must be segmented. A two-stage segmentation is shown in Fig. 3.16. Registers must be inserted in all positions where a connection crosses the dotted line. Thus, seven registers must be added. Assuming that the propagation time of every part of a segmented connection is still equal to $t_{connection}$, the following condition must hold:

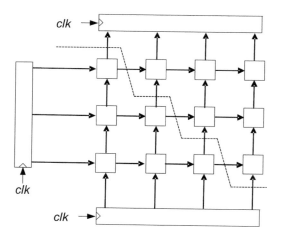

Fig. 3.16 Two-stage segmentation

Fig. 3.15 Combinational circuit

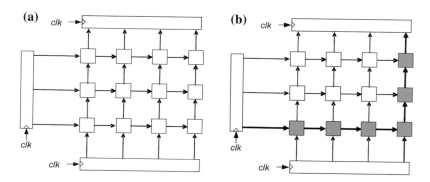

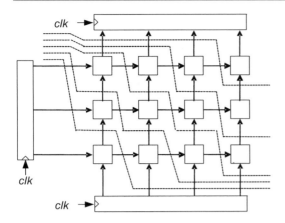

Fig. 3.17 Five-stage segmentation

$$T_{CLK} > 3 \cdot t_{cell} + 4 \cdot t_{connection} + t_{SU} + t_P. \tag{3.16}$$

A five-stage segmentation is shown in Fig. 3.17. In this case, 32 registers must be added and the following condition must hold:

$$T_{CLK} > t_{cell} + 2 \cdot t_{connection} + t_{SU} + t_P. \tag{3.17}$$

Consider a concrete example.

Example 3.3 Implement a 128-bit adder made up of four 32-bit adder components. A combinational implementation is described in Fig. 3.18. In this case, the connection propagation times are negligible with respect to the component delays. Thus, the computation time T of the circuit is equal to 4 · T_{adder}, where T_{adder} stands for the computation time of a 32-bit adder. It corresponds to the critical path from the c_{in} input to the c_{out} output.

A four-stage segmentation is shown in Fig. 3.19. Every stage includes one 32-bit adder

so that the minimum clock cycle, as well as the minimum time interval between successive data inputs, is equal to T_{adder}. The corresponding circuit is shown in Fig. 3.20. In total (7 · 32 + 1) + (6 · 32 + 1) + (5 · 32 + 1) = 579 additional flip-flops are necessary in order to separate the pipeline stages.

In order to implement this circuit, a predefined carry-select adder (Parhami 2000; Deschamps et al. 2012) is used to synthesize the 32-bit adder component. A parameterized VHDL model *carry_select_adder.vhd* is available at the Authors' web site. The parameters m and k define the structure of the carry-save adder: it consists of m k-bit adders. In this case, the chosen parameter values are $m = 4$ and $k = 8$ so that the total number of bits is 4 · 8 = 32. A VHDL model *adder128pipeline.vhd* of the circuit of Fig. 3.20 is available at the Authors' web site. Part of the simulation result is shown in Fig. 3.21. The circuit computes

0123456789abcdef0123456789abcdee + fedcba9876543210fedcba9876543210 + 1 = 0ffffffffffffffffffffffffffffffff,
11111111111111111111111111111110 + 00000000000000000000000000000000 + 1 = 01111111111111111111111111111111,
99999999999999999999999999999999 + 88888888888888888888888888888888 + 1 = 12222222222222222222222222222222,
bbbbbbbbbbbbbbbbbbbbbbbbbbbbbbbb + 44444444444444444444444444444444 + 1 = 1000000000000000000000000000000000
(All numbers in hexadecimal).

The latency is equal to 4 cycles, and the pipeline rate is equal to 1 cycle.

Fig. 3.18 Twelve 8-bit adder

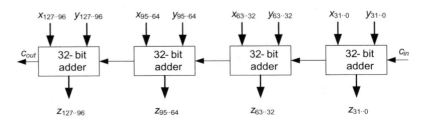

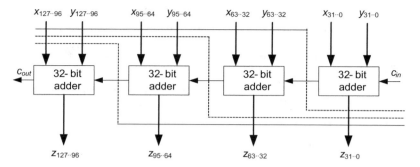

Fig. 3.19 Four-stage segmentation

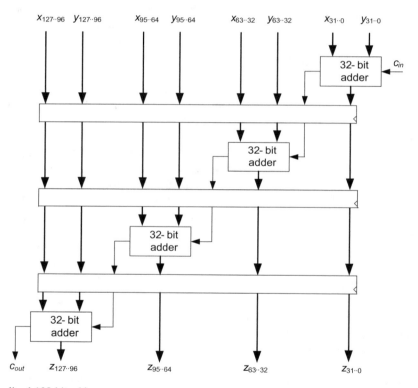

Fig. 3.20 Pipelined 128-bit adder

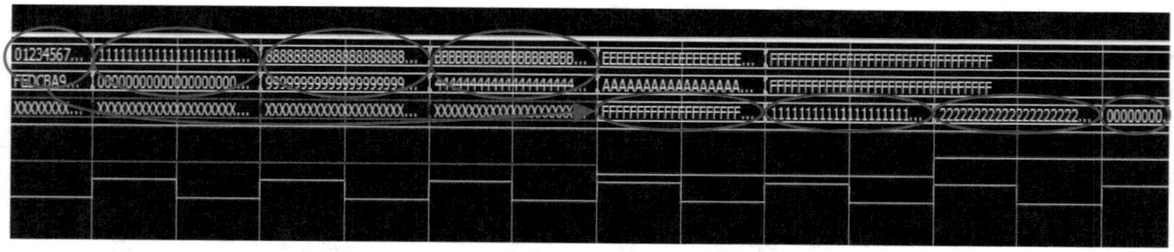

Fig. 3.21 Simulation of a pipelined 128-bit adder (courtesy of Mentor Graphics)

3.3.2 Digital Signal Processing

A digital signal s is a sequence of successive numbers $x_0, x_1, x_2 \ldots$ where

- x_i is a signal value belonging to some pre-defined discrete type (an integer, a fixed-point number, a floating-point number).
- x_i is the value of s at time $i \cdot T$ where T is a predefined sample period.

Generally, digital signals are the encoded representation of analog signals. As an example, in order to store an analog signal on a digital support, for example a digital video disk (DVD), the analog signal value is sampled every T seconds and every sample is converted to a fixed-point number. Those operations are performed by a component called analog-to-digital converter (ADC). According to the Shannon–Nyquist theorem, if the maximum frequency of the analog signal is equal to f_{max} hertz and if the sample frequency $1/T$ is greater than $2 \cdot f_{max}$, then the original analog signal can be reconstructed by means of a component called digital-to-analog converter (DAC). The reconstructed analog signal contains an additional noise due to quantization errors, but the noise amplitude can be minimized—practically eliminated—by encoding the signal samples with a sufficient number of bits.

Operations such as filtering, storing, compressing, ciphering of analog signals are complex. Digital signal processing systems use the possibility of converting analog signals to digital signals, without practically any loss of information, in order to execute those complex operations with digital circuits: the analog signal is first converted to a digital signal; then it is processed by means of a digital circuit; finally, the resulting digital signal is converted to an analog signal. A simplified block diagram of the whole system is shown in Fig. 3.22.

The data input of the digital processing block of Fig. 3.22 is a sequence of successive numbers $x = x_0, x_1, x_2 \ldots$ where x_i is the quantized value of the analog input signal s_x at time $i \cdot T$, being T the sample period: $x_i \cong s_x(i \cdot T)$. The data output of the digital processing block is also a sequence of successive numbers $y = y_0, y_1, y_2, \ldots$ where y_i is the value of the analog output signal s_y at time $i \cdot T$. Thus,

$$s_y(i \cdot T) = y_i = F(x_0, x_1, x_2 \cdots, x_i)$$
$$\cong F(s_x(0), s_x(T), s_x(2 \cdot T) \cdots, s_x(i \cdot T))$$
$$(3.18)$$

being F a function that describes the behavior of the digital processing block.

As both the input and output sequences x and y are synchronized by the same sampling signal, the throughput is a central characteristic of this type of circuit: a new sample x_i is inputted every T time units, and a processed value y_i must be outputted every T time units. On the other hand, the latency is not a so important issue as the implemented processes (filters and so on) generally are time-invariant processes: a delay on the input signal only causes an equal delay of the output signal.

To conclude, due to the central importance of throughput, digital signal processing circuits are clear candidates at pipelined implementations.

To illustrate the preceding ideas, consider a common type of digital signal processing circuit:

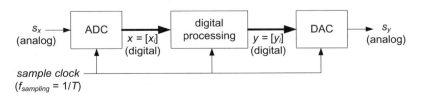

Fig. 3.22 Digital processing of signals

a digital filter. There are many textbooks on digital filters, for example (Smith 2007). Consider the system of Fig. 3.22 assuming that s_x is an analog signal, for example an audio signal, and that s_y must be a signal similar to s_x but with a different frequency spectrum: part of the frequencies are passed unaltered, while other are attenuated. Typical examples are low-pass filters, high-pass filters, band-pass filters and band-stop filters.

In the case of a digital filter, the relation between $x = [x_i]$ and $y = [y_i]$ that are the input and output signals of the digital processing circuit of Fig. 3.22 is an equation of the following type

$$y_i = a_0 \cdot x_i + a_1 \cdot x_{i-1} + \ldots + a_k \cdot x_{i-k} - b_1 \cdot y_{i-1} - b_2 \cdot y_{i-2} - \ldots - b_l \cdot y_{i-1},$$

$$(3.19)$$

being $\{a_0, a_1, \ldots, a_k, b_1, b_2, \ldots, b_l\}$ a set of predefined constants that belong to some previously defined type. The order of the filter is the larger of k and l.

Only a particular type of filter will be considered in this example: if all coefficients b_i are equal to zero, so that

$$y_i = a_0 \cdot x_i + a_1 \cdot x_{i-1} + \cdots + a_k \cdot x_{i-k}, \quad (3.20)$$

the obtained circuit is a finite impulse response (FIR) filter. If the input signal is a one-period impulse at time $p \cdot T$, that is a signal $[x_i]$ defined as follows

$$x_p = 1 \text{ and } x_i = 0 \; \forall i \neq p, \quad (3.21)$$

then the output signal is a signal $[h_i]$ where

$$h_i = 0 \text{ if } i < p, h_p = a_0, h_{p+1} = a_1, \ldots, h_{p+k} = a_k, h_i = 0 \text{ if } i > p+k.$$

$$(3.22)$$

(3.22) is a direct consequence of (3.20) and (3.21).

The block diagram of a circuit that implements Eq. (3.20) is shown in Fig. 3.23: it consists of a set of k registers that store $x_{i-1}, x_{i-2}, \ldots, x_{i-k}$ and a combinational circuit that computes (3.20).

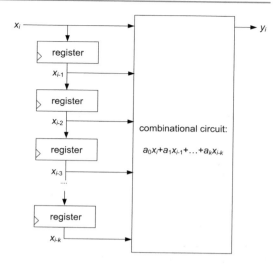

Fig. 3.23 FIR filter structure

The latency is approximately equal to the delay T_{cc} of the combinational circuit, and the throughput is equal to $1/latency \cong 1/T_{cc}$.

There are several ways to synthesize the combinational circuit. In order to minimize the latency and to maximize the throughput, the circuit can be implemented as follows: a set of $k + 1$ multipliers that compute in parallel all products $a_j \cdot x_{i-j}$ (multipliers by a constant) and a fast $(k + 1)$-operand adder (Fig. 3.24a). Assume that all data (inputs and outputs) can be expressed as m-bit 2's complement fixed-point numbers and that a tree of two-operand adders is used (Chap. 8 of Parhami 2000 and Chap. 11 of Deschamps et al. 2006). Then, the magnitude order of the multi-operand adder delay is $O(m + \log_2 k)$. However, a more regular and expandable structure is generally preferred (Fig. 3.24b): it consists of $k+1$ multiplier-accumulator components (MACs) that compute $acc + a_j \cdot x_{i-j}$ (Fig. 3.24c).

If the two-operand adder included within each MAC consists of m full adders (an m-bit ripple adder), then the set of all two-operand adders is an array of $m \cdot k$ full adders and the magnitude order of the corresponding multi-operand adder delay is $O(m + k)$. An example with $m = 4$ and $k = 3$ is shown in Fig. 3.25: the total delay of the adder array is equal to $(m + k) \cdot T_{FA}$; one of the critical paths is highlighted in Fig. 3.25. The latency of the complete combinational circuit is equal to $T_{mult(m)} + (m + k) \cdot T_{FA}$ where $T_{mult(m)}$ is the delay of a multiplier by a constant whose output is an m-bit 2's complement fixed-point number.

So, the circuit of Fig. 3.24b is regular and easily expandable, but has a longer latency and a lower throughput than a circuit using a fast multi-operand adder ($O(m + k)$ vs. $O(m + \log_2 k)$).

To increment the throughput, the solution is pipelining. An FIR filter using a pipelined implementation of the circuit of Fig. 3.24b is shown in Fig. 3.26a. Its latency is equal to $(k + 1) \cdot T_{clk}$,

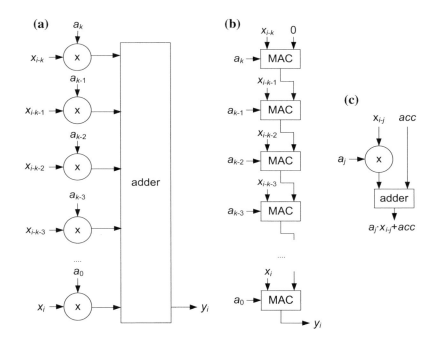

Fig. 3.24 **a** Fast multi-operand adder. **b** Multiplier-Accumulator Components (MAC)

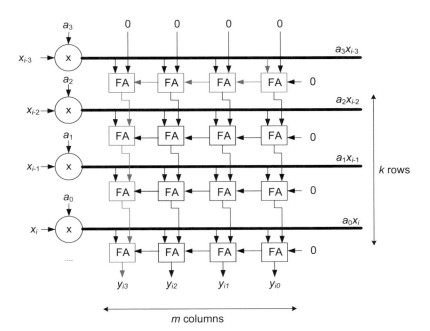

Fig. 3.25 Example of combinational circuit with $m = 4$ and $k = 3$

Fig. 3.26 Pipelined FIR
filters: **a** First version.
b Equivalent circuit

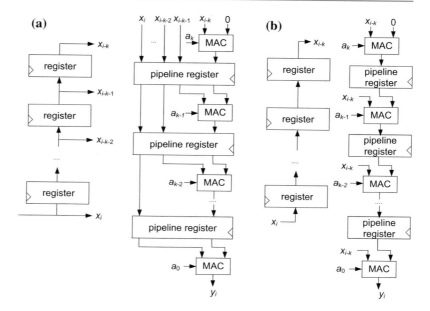

where T_{clk} must be greater than the delay T_{MAC} of a multiplier-accumulator component (MAC), and the maximum throughput is approximately equal to $1/T_{MAC}$.

Observe that the inputs x_i, x_{i-1}, ..., x_{i-k+1} of the non-pipelined combinational circuit (Fig. 3.24b) now go through k, $k-1$, $k-2$, ... pipeline registers, so that all right inputs of the MAC actually are equal to x_{i-k}. So, the circuit of Fig. 3.26a is equivalent to that of Fig. 3.26b: a set of k registers that generate x_{i-k} and a $(k+1)$-stage pipeline consisting of $k+1$ MAC. The k registers that generate x_{i-k} are a sort of delay line inserted on the input size of the digital filter whose only (useless) effect is to increment the latency. Obviously, those registers can be removed. In this way, the circuit of Fig. 3.27 is obtained.

Check the behavior of the circuit of Fig. 3.27: it is defined by the following equation:

$$y_i = a_0 \cdot x_i + \ldots + \left(a_{k-2} \cdot x_i\right)^{d^{k-2}}$$
$$+ \left(a_{k-1} \cdot x_i\right)^{d^{k-1}} + \left(a_k \cdot x_i\right)^{d^k} \quad (3.23)$$

where z^d stands for z delayed by a clock period. Then, as a_1, a_2, ... are constants,

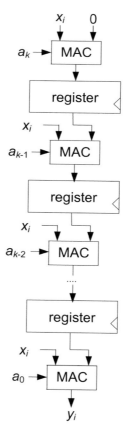

Fig. 3.27 FIR filter

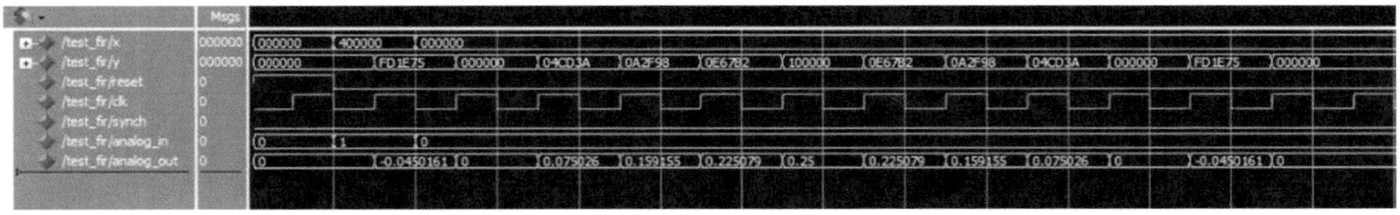

Fig. 3.28 Impulse response

$$\left(a_j \cdot x_i\right)^{d^j} = a_j \cdot x_{i-j}. \tag{3.24}$$

Thus, according to (3.23) and (3.24)

$$y_i = a_0 \cdot x_i + \cdots + a_{k-2} \cdot x_{i-k-2} + a_{k-1} \\ \cdot x_{i-k-1} + a_k \cdot x_{i-k}, \tag{3.25}$$

that is the equation that defines an FIR filter (3.20).

The circuit of Fig. 3.27 has the same maximum throughput (approximately $1/T_{MAC}$) as the circuit of Fig. 3.26a, but its latency is equal to one clock period instead of $k + 1$.

A complete VHDL model *fir_filter.vhd* is available at the Authors' web site. The filter parameters k, m, a_0, a_1, ..., a_k are defined within a package *fir_parameters*. All data d (x_i, y_i, a_i) are real numbers. They are assumed to belong to the interval

$$-2 \le d < 2$$

and are represented under the form $d = D \cdot 2^{-m + 2}$ where D is an m-bit 2s complement integer, so that—$2^{m-1} \le D < 2^{m - 1}$ and thus— $2 \le D \cdot 2^{-m + 2} < 2$. In other words, all data are m-bit 2s complement fixed-point numbers with $m - 2$ fractional bits:

$$d_{m-1} \, d_{m-2} \, d_{m-3} \, d_{m-4} \cdots d_0.$$

Within the VHDL model, the exponent 2^{-m+2} is implicit so that only the significant D (an m-bit 2s complement integer) is explicitly defined.

To test the VHDL model, a particular low-pass filter with the following parameter values

$$k = 10, m = 24, a_0 = a_{10} = -0.045016,$$
$$a_1 = a_9 = 0, a_2 = a_8 = 0.075026,$$
$$a_3 = a_7 = 0.159155, a_4 = a_6 = 0.225079,$$
$$a_5 = 0.25,$$

has been implemented.

The response to a one-period impulse is shown in Fig. 3.28: the input signal is [..., 0, 1, 0, ...], and the output signal is [... 0, a_0, a_1, ..., a_k, 0, ...] that is signal [h_i] defined by (3.22).

The response to a periodic signal [... −1, 1, −1, 1, −1, 1, ...], with a period equal to twice the sample period, is shown in Fig. 3.29. The output signal is a periodic symmetric signal with very small maximum and minimum values (±0.058132) that correspond to the following sums of coefficients:

$$- a_0 + a_1 - a_2 + a_3 - a_4 + a_5 - a_6 + a_7 - a_8 \\ + a_9 - a_{10} = 0.058132,$$
$$a_0 - a_1 + a_2 - a_3 + a_4 - a_5 + a_6 - a_7 + a_8 \\ - a_9 + a_{10} = -0.058132.$$

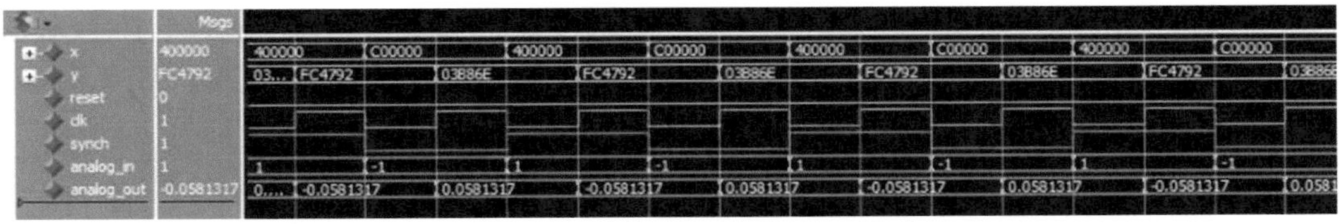

Fig. 3.29 Response to a periodic input signal

3.4 Interconnection of Pipelined Components

Assume that several pipelined circuits are used as computational resources (predefined components) for generating a new pipelined circuit. For example, consider the two following pipelined components:

- a two-stage adder that computes $x + y$ in 2 clock periods ($2 \cdot T$ time units) with a pipeline rate equal to $1/T$;
- a three-stage multiplier that computes $x \cdot y$ in 3 clock periods ($3 \cdot T$ time units) with a pipeline rate equal to $1/T$.

The corresponding symbols are shown in Fig. 3.30.

When interconnecting pipelined components, it may be necessary to add registers to maintain the correct synchronization of the processed data. As an example, the two previously defined pipelined components can be used to implement a circuit that computes two functions $g = a + b$ and $f = (a + b) \cdot c + d$ according to the schedule of Fig. 3.31a. The corresponding circuit block diagram is shown in Fig. 3.31b. In order to synchronize the input c with the output $a + b$ of the first adder, two (so-called) skewing registers are inserted. Similarly, five skewing registers are added to synchronize the input d with the multiplier output $(a + b) \cdot c$. On the output side, five

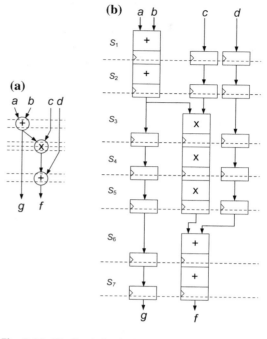

Fig. 3.31 Pipelined circuit that computes $g = a + b$ and $f = (a + b) \cdot c + d$

deskewing registers must be added to synchronize the output g with the output f of the second adder. In this way, the obtained circuit (Fig. 3.31b) has seven pipeline stages (S_1 to S_7). The latency is equal to $7 \cdot T$, and the pipeline rate is equal to $1/T$.

3.5 Self-timed Circuits

A pipelined circuit consists of a set of stages that work in parallel. All stages transfer their output data to the next stage under the control of a common synchronization command (e.g., en_pipe in Fig. 3.13). When all stages have approximately the same computation time, this is an efficient option. In fact, this characteristic (almost equal delays) is one of the aspects to be taken into account when segmenting a circuit. However, it can happen that the stage delays are data-dependent so that in some cases the computation time could be relatively short and in other cases it could be relatively long. Then, the pipeline period should be greater than the longest

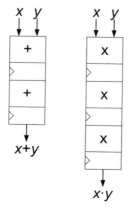

Fig. 3.30 Two pipelined components

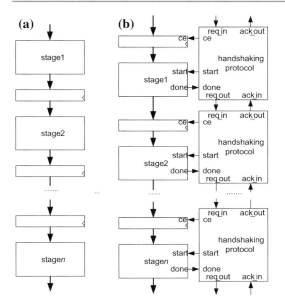

Fig. 3.32 a Pipelined circuit. **b** Self-timed pipelined circuit

of all delays, probably an inefficient solution. An alternative solution is *self-timing*.

As a generic example, consider the pipelined circuit of Fig. 3.32a. To each stage, for example number *i*, are associated a maximum delay $t_{MAX}(i)$ and an average one $t_{AV}(i)$. The minimum time interval between successive data inputs is

$$\delta = max\{t_{MAX}(1), t_{MAX}(2), \ldots, t_{MAX}(n)\}, \tag{3.26}$$

and the circuit latency *T* is

$$T = n \cdot max\{t_{MAX}(1), t_{MAX}(2), \ldots, t_{MAX}(n)\}. \tag{3.27}$$

A self-timed version of the same circuit is shown in Fig. 3.32b. The control is based on a *Request/Acknowledge* handshaking protocol. An example of handshaking protocol is shown in Fig. 3.33. It works as follows:

- When stage *i*-1 has completed a computation, it raises the *req_out* output signal to inform stage *i* that processed data is available.

- As soon as stage *i* is idle—that means that it has completed its current computation and has received from stage *i* + 1 an acknowledge signal asserting that the latest processed data generated by stage *i* has been registered within stage *i* + 1—the input data generated by stage *i* − 1 is registered (*ce* = 1), and an *ack_out* signal is issued to stage *i* − 1.
- The *start* signal of stage *i* is raised; after some amount of time, the *done* signal of stage *i* goes high indicating the completion of the computation.
- A *req_out* signal to stage *i* + 1 is issued by stage *i*; when stage *i* + 1 is idle, the output of stage *i* is registered within stage *i* + 1 and an *ack_out* signal to stage *i* is issued; and so on.

If the probability distribution of the internal data is uniform, inequalities (3.26) and (3.27) can be substituted by the following ones:

$$\delta = max\{t_{AV}(1), t_{AV}(2), \ldots, t_{AV}(n)\}, \tag{3.28}$$
$$T = t_{AV}(1) + t_{AV}(2) + \ldots + t_{AV}(n). \tag{3.29}$$

The protocol of Fig. 3.33 is implemented by the finite-state machine of Fig. 3.34. A VHDL model *protocol.vhd* is available at the Authors' web site.

Example 3.4 A self-timed version of the distance computation circuit of Fig. 3.13 is shown in Fig. 3.35. It consists of the following components: *step*1 (Fig. 3.36), *step*2 (Fig. 3.37), two handshaking protocol circuits, three registers and a simplified handshaking circuit that controls the output register.

A VHDL model *distanceST.vhd* is available at the Authors' web site. Part of the simulation result is shown in Fig. 3.38. The circuit computes

$$\lfloor \left[|3 - 7|^2 + |18 - 6|^2 + |1 - 5|^2 \right]^{1/2} \rfloor = 13,$$

$$\lfloor \left[|11 - 7|^2 + |10 - 38|^2 + |9 - 5|^2 \right]^{1/2} \rfloor = 28,$$

$$\lfloor \left[|0 - 255|^2 + |0 - 255|^2 + |0 - 255|^2 \right]^{1/2} \rfloor = 441.$$

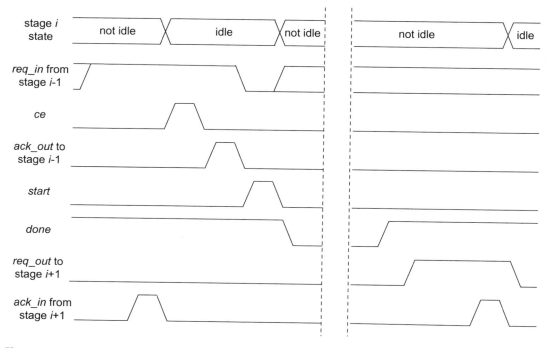

Fig. 3.33 Handshaking protocol

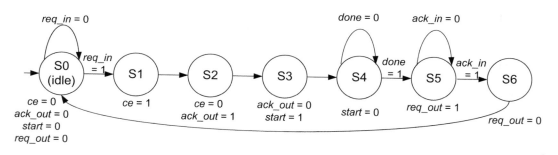

Fig. 3.34 Handshaking protocol control state machine

In the former example, the delays of the used components are not data-dependent, so that a self-timed circuit has no real advantage over the pipelined circuit of Fig. 3.13. Actually, the input rate is practically the same and the latency would also be practically the same without the input register.

Self-timed implementations can be considered even in the case of combinational circuits. The problem is the generation of the *done* signal. For that, an interesting method consists in using a redundant encoding of the binary signals (Sect. 10.4 of Rabaey et al. 2003): every signal s is represented by a pair (s_1, s_0) according to the definition of Table 3.3.

The circuit must have an additional *reset* input and is designed in such a way that during the initialization (*reset* = 1) and as long as the value

Table 3.3 Redundant encoding

s	s_1	s_0
Reset or *in transition*	0	0
0	0	1
1	1	0

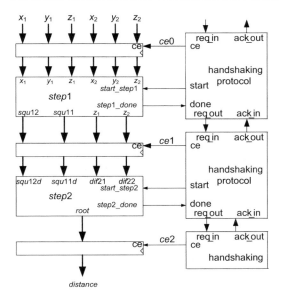

Fig. 3.35 Self-timed distance computation circuit

of s has not yet been computed, the value of the pair (s_1, s_0) that encodes s is $(0, 0)$. Once the value of s is known, $s_1 = s$ and $s_0 = not(s)$.

Assume that the circuit includes n signals s_1, s_2, ..., s_n. Every signal s_i is substituted by a pair (s_{i1}, s_{i0}). Then, the *done* flag is computed as follows:

$$done = (s_{11} + s_{10}) \cdot (s_{21} + s_{20}) \ldots (s_{n1} + s_{n0}).$$

During the initialization (*reset* = 1) and as long as at least one of the signals is in transition, the corresponding pair is equal to $(0, 0)$, so that *done* = 0. The *done* flag will be raised only when all signals have a stable value.

In the following example, only the signals belonging to the critical path of the circuit are encoded.

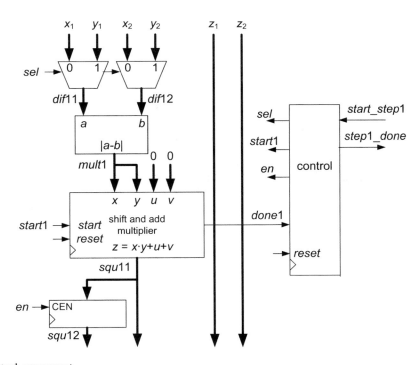

Fig. 3.36 *step*1 component

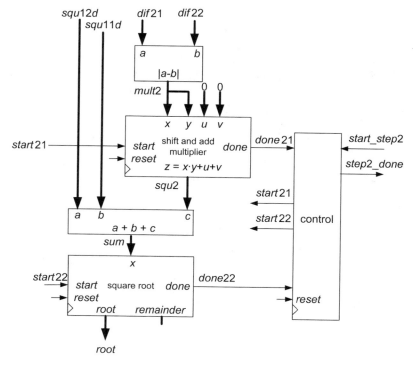

Fig. 3.37 *step*2 component

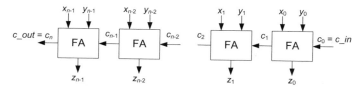

Fig. 3.38 Simulation of *distanceST.vhd* (courtesy of Mentor Graphics)

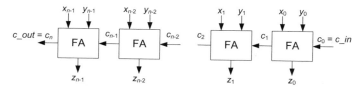

Fig. 3.39 Ripple carry adder

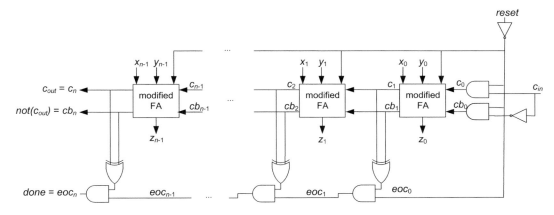

Fig. 3.40 Ripple carry adder with end of computation detection

Example 3.5 Generate an *n*-bit ripple carry adder (Fig. 3.39) with end of computation detection.

For that, all signals belonging to the carry chain, that is c_0, c_1, c_2, ..., c_n, are represented under the form $(c_0, cb_0), (c_1, cb_1), (c_2, cb_2), ..., (c_n, cb_n)$. The circuit is shown in Fig. 3.40. The modified FA cell implements the following equations:

$$c_{i+1} = (x_i \cdot y_i + x_i \cdot c_i + y_i \cdot c_i) \cdot \overline{reset}, cb_{i+1}$$
$$= (\overline{x_i} \cdot \overline{y_i} + \overline{x_i} \cdot \overline{c_i} + \overline{y_i} \cdot \overline{c_i}) \cdot \overline{reset},$$
$$(3.30)$$

$$z_i = x_i \oplus y_i \oplus c_i \qquad (3.31)$$

During the initialization (*reset* = 1), c_i and cb_i are equal to 0, $\forall i \in \{0, 1, ..., n\}$. When *reset* goes down

$$c_0 = c_{in}, \quad cb_0 = \overline{c_{in}}, \qquad (3.32)$$

and the circuit starts computing

$$c_{i+1} = (x_i \cdot y_i + x_i \cdot c_i + y_i \cdot c_i),$$
$$cb_{i+1} = (\overline{x_i} \cdot \overline{y_i} + \overline{x_i} \cdot \overline{c_i} + \overline{y_i} \cdot \overline{c_i}), \qquad (3.33)$$

for *i* = 0, 1, 2, ... starting from the least significant bits c_1 and cb_1 up to the most significant bits c_n and cb_n. The end of computation is detected when

$$cb_i = \overline{c_i}, \forall i \in \{0, 1, ..., n\}. \qquad (3.34)$$

A complete and synthesizable VHDL model *adder_ST2.vhd* of the circuit of Fig. 3.40 is available at the Authors' web site. A test file *test_adder_ST2.vhd* is also available. In order to observe the carry chain delay, *after* clauses have been added (1 ns for c_{i+1} and cb_{i+1}, 0.2 ns for eoc_{i+1}). For synthesis purpose, they must be deleted.

3.6   Exercises

1. Generate VHDL models of different pipelined 128-bit adders.
2. In the following combinational circuits, the delays of every cell and of every connection are equal to 5 and 2 ns, respectively.

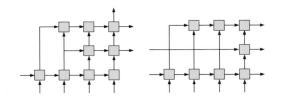

For each circuit:

a. Compute the combinational delay.
b. Segment the circuit in two stages. How many registers must be added?
c. Segment the circuit in three stages. How many registers must be added?
d. What is the maximum number of segmentation stages?

e. Assume that the cutting of a connection generates two new connections whose delays are still equal to 2 ns, and that the registers have a propagation delay of 1 ns and a setup time of 0.5 ns. What is the maximum frequency of circuits b. and c.?

Bibliography

De Micheli G (1994) Synthesis and Optimization of Digital Circuits. McGraw-Hill, New York

Deschamps JP, Sutter G, Cantó E (2012) Guide to FPGA Implementation of Arithmetic Functions. Springer, Dordrecht

Deschamps JP, Bioul G, Sutter G (2006) Synthesis of Arithmetic Circuits. Wiley, Hoboken, New Jersey

Parhami B (2000) Computer Arithmetic: Algorithms and Hardware Design. Oxford University Press, New York

Rabaey JM, Chandrakasan A, Nikolic B (2003) Digital Integrated Circuits: a Design Perspective. Prentice Hall, Upper Saddle River

Smith JO (2007) Introduction to Digital Filters. W3K Publishing

Loops are present in practically any algorithm so that their optimal implementation is a basic aspect of the synthesis of digital circuits. Iterative and sequential implementations are considered and synthesis techniques such as loop-unrolling and digit-serial processing are presented.

4.1 Introductory Example

The introductory example of Chap. 1 (base-2 logarithm) is revisited with two simple but instructing modifications. The original algorithm (Algorithm 1.1) is the following.

Algorithm 1.1 Base-2 logarithm

```
z = x; i = p;
while i > 0 loop
  z = z²;
  if z ≥ 2 then y_{i-p-1} = 1; z = z/2;
  else y_{i-p-1} = 0;
  end if;
  i = i-1;
end loop;
```

The corresponding data path and control unit are shown in Figs. 1.1 and 1.2 of Chap. 1. According to this algorithm, register z is updated two times during the loop body execution:

- First, the value stored in register z is squared and the result is stored within the same register z ($z = z^2$);
- Then, if the value stored in register z is greater than or equal to 2, this value is divided by 2 and the result is stored within the same register z ($z = z/2$).

To execute the second operation, the control unit reads the value stored in register z (actually the most significant bit), and in function of the read value, it generates the following control signal values:

$load\_z = 1$ and $sel\_z = 2$ if $z \geq 2$, $load\_z = 0$ and $sel\_z = don't\ care$ if $z < 2$.

Consider now a slightly modified version of Algorithm 1.1:

Algorithm 4.1 Base-2 logarithm (modified version)

```
z = x; i = p;
while i > 0 loop
  if z² ≥ 2 then y_{i-p-1} = 1; z = z²/2;
  else y_{i-p-1} = 0; z = z²;
  end if;
  i = i-1;
end loop;
```

The difference is that the squaring of z is included within a branching condition (if $z^2 \geq 2$ then $\cdots$).

In order to implement this conditional instruction, a straightforward solution is to design a computation resource able to execute in one cycle the following operation

```
if z² ≥ 2 then t = 1; next_z = z²/2;
else t = 0; next_z = z²;
end if;
```

where t is a binary value (flag, condition variable). An example of implementation is given in Fig. 4.1.

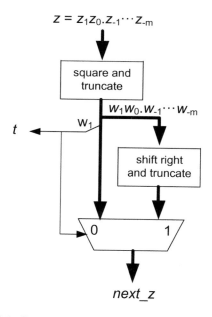

Fig. 4.1 Computation resource

Then, the circuit of Fig. 1.1 can be replaced by the data path of Fig. 4.2. Observe that the selection of the value to be stored within register z (either z^2 or $z^2/2$) is no longer defined by the control unit; it is internally performed by a multiplexer within the computation resource (Fig. 4.1) of the data path. This is an example of how a part of the control task can be moved from the control unit to the data path.

Another modification is the way in which the end of the loop execution is detected: instead of checking whether the value stored in register i is equal to 0 or not, the output value of the circuit that computes $i - 1$ is considered. This permits to update register i ($i = i - 1$) at the same time as register z ($z = next\_z$) and register y (*shift y*). The control unit is shown in Fig. 4.3. When in state 2, the branching to either state 2 or state 0 is based on the next value ($i - 1$) of register i instead of the current value. This is a kind of control anticipation.

In conclusion, in this example two simple circuit modifications (moving of part of the control to the data path and control anticipation) permit to reduce the number of cycles associated with the loop body execution of the algorithm.

A VHDL model *logarithm_circuit_bis.vhd* is available at the Authors' web site.

The implementation of Figs. 4.2 and 4.3 is based on the fact that Algorithm 4.1 mainly consists of a loop that is executed p times. The computation resource of Fig. 4.1 is (re)used p times to compute the values of z and y_{i-p-1}. The

Fig. 4.2 Data path

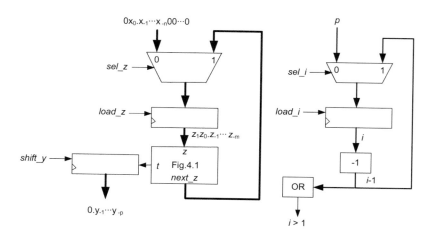

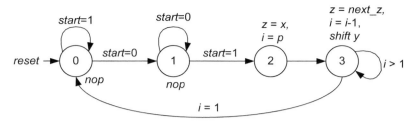

Fig. 4.3 Control unit

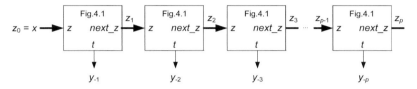

Fig. 4.4 Iterative circuit

successive values of z are stored within a parallel register, and the values of y_{-1}, y_{-2}, ... , y_{-p} are stored within a shift register. To complete the circuit a p-state counter, several controllable connections (multiplexers) and a control unit must be added. The computation time is equal to $p \cdot T_{clk}$ where the clock period T_{clk} must be greater than the computation time of the data path of Fig. 4.2, including the propagation time, hold time and setup time of the registers.

Another way to implement Algorithm 4.1 is an iterative circuit. For that consider the following equivalent Algorithm:

Algorithm 4.2 Base-2 logarithm (modified version)

```
z₀ = x;
for i = 1 to p loop
  if z²ᵢ₋₁ ≥ 2 then y₋ᵢ = 1; zᵢ = z²ᵢ₋₁/2;
  else y₋ᵢ = 0; zᵢ = z²ᵢ₋₁;
  end if;
end loop;
```

In this case, p copies of the computation resource of Fig. 4.1 must be used, one for every value of index i. The resulting circuit is shown in Fig. 4.4.

The computation time is equal to $p \cdot T_{resource}$ where $T_{resource}$ is the delay of the combinational circuit of Fig. 4.1. As the data path of Fig. 4.2 is made up of the combinational circuit of Fig. 4.1 plus registers and connections, the clock period T_{clk} of the circuit of Figs. 4.2 and 4.3 must be greater that the delay of the circuit of Fig. 4.1. Thus, $p \cdot T_{resource} < p \cdot T_{clk}$ and the iterative implementation (Fig. 4.4) is faster than the sequential implementation (Figs. 4.2 and 4.3). On the other hand, the cost of the iterative circuit is equal to $p \cdot C_{resource}$ where $C_{resource}$ is the cost of the combinational circuit of Fig. 4.1 while the cost of the sequential circuit is equal to $C_{resource} + C_{multiplexers} + C_{registers} + C_{counter} + C_{control}$. Unless p is very small, the sequential implementation is cheaper.

A VHDL model *logarithm_circuit_iterative. vhd* of the circuit of Fig. 4.4 is available at the Authors' web site.

A rather rough conclusion might be that iterative circuits are faster and sequential circuits are most cost-effective.

Intermediate options could also be contemplated. Instead of completely unroll Algorithm 4.1 so as to get Algorithm 4.2, consider the following equivalent algorithm.

Algorithm 4.3 Base-2 logarithm (modified version)

```
--p even equal to 2k
z₀ = x; i = p/2;
while i > 0 loop
  if z₀² ≥ 2 then y₂ᵢ₋ₚ₋₁ = 1; z₁ = z₀²/2;
  else y₂ᵢ₋ₚ₋₁ = 0; z₁ = z₀²;
  end if;
  if z₁² ≥ 2 then y₂ᵢ₋ₚ₋₂ = 1; z₀ = z₁²/2;
  else y₂ᵢ₋ₚ₋₂ = 0; z₀ = z₁²;
  end if;
  i = i-1;
end loop;
```

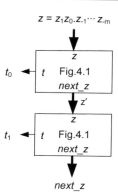

$$z = z_1 z_0 . z_{-1} \cdots z_{-m}$$

Fig. 4.5 Computation resource

Every loop body execution computes the value of two successive bits y_{i-p-1} and y_{i-p-2} of y. A straightforward solution is to design a computation resource able to execute in one cycle the following operations

```
if z² ≥ 2 then t₀ = 1; z' = z²/2;
else t₀ = 0; z' = z²; end if;
if z'² ≥ 2 then t₁ = 1; next_z = z'²/2;
else t₁ = 0; next_z = z'²; end if;
```

where t_0 and t_1 are binary values. An example of implementation is given in Fig. 4.5: it consists of two serially connected copies of the circuit of Fig. 4.1.

The data path of the complete circuit is shown in Fig. 4.6, and the control unit is the same as before (Fig. 4.3). A VHDL model *logarithm_circuit_ter.vhd* of the circuit is available at the Authors' web site.

The computation time is equal to $(p/2) \cdot T_{clk}$ where the clock period T_{clk} must be greater than the computation time of the data path of Fig. 4.5, including the propagation time, hold time and setup time of the registers. The cost is equal to $2 \cdot C_{resource} + C_{multiplexers} + C_{registers} + C_{counter} + C_{control}$ where $C_{resource}$ is the cost of the combinational circuit of Fig. 4.1.

Fig. 4.6 Data path

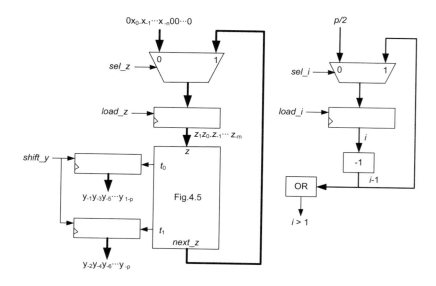

4.2   Iterative Versus Sequential Implementation of Loops

Some general considerations about loop implementations are now presented. The following generic iterative algorithm mainly consists of a loop instruction whose loop body is a procedure *operations (a: in; b: out)* with an input parameter *a* and an output parameter *b*.

Algorithm 4.4 Generic iterative algorithm

```
data₀ := initial_values;
for i in 1 .. p loop
  operations(data_{i-1}, data_i);
end loop;
final_results := data_p;
```

Assume that a component that implements the procedure *operations* has been previously developed. It might be either a combinational circuit with a delay equal to $T_{operations}$ or a sequential component with a computation time equal to N clock cycles, in which case $T_{operations} = N \cdot t_{clock}$ where t_{clock} is the period of the clock signal internally used by the component *operations*. Two straightforward implementations of the algorithm are shown in Fig. 4.7. The first one (Fig. 4.7a) is an iterative implementation: the loop construct has been completely expanded (unrolled), and the corresponding circuit uses p instances of the component *operations*. Its cost and delay are

$$C_{iterative} = p \cdot C_{operations}, T_{iterative} \leq p \cdot T_{operations}$$
(4.1)

where $C_{operations}$ and $T_{operations}$ are the cost and the delay of the component that implement the procedure *operations*. If *operation* is a combinational component, the delay could be smaller than $p \cdot T_{operations}$ as the critical paths of the complete p-component circuit do not necessarily consist of concatenated critical paths of the individual components. For example, the delay of an n-bit ripple carry adder is proportional to n; nevertheless, the delay of two serially connected ripple carry adders that compute $(a + b) + c$ is proportional to $n + 1$, not to $2n$.

The second one (Fig. 4.7b) is a sequential implementation using only one instance of the component *operations*. Its main characteristics are

$$C_{sequential} = C_{operations} + C_{extra}, T_{sequentiall}$$
$$= p \cdot T_{clk} \text{ with } T_{clk} > T_{operations} + T_{extra}$$
(4.2)

where

- C_{extra} is the cost of the additional circuits (registers, multiplexers and control) necessary to execute a p-step iteration with only one component. In particular, it includes the circuit (e.g., a mod p counter) that controls the number of executions of the loop body.
- T_{extra} is the additional delay of the critical paths of the data path due to the additional registers and connections.

In spite of the chosen names (iterative vs. sequential) both implementations may be considered as being iterative, the first one over the space domain (silicon surface and printed circuit board area) and the second over the time domain (time multiplexing of operations).

To compare (4.1) and (4.2), it is assumed that C_{extra} does not depend on p. This is not completely correct as one of the functions of the control circuit is to count the number of

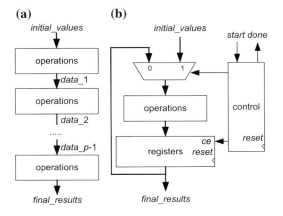

(a)

initial_values

operations

data_1

operations

data_2

.....

data_p-1

operations

final_results

(b)

initial_values

start done

0   1

operations   control

reset

ce

registers   reset

final_results

Fig. 4.7 Iterative algorithm implementation: **a** Unrolled implementation. **b** Sequential implementation

executions of the loop body. Nevertheless, this is only a part of the additional circuit and its cost is proportional to $log_2 p$, not to p. Thus, according to (4.1) and (4.2), the sequential implementation (Fig. 4.7b) has a lower cost than the iterative implementation (Fig. 4.7a) if $p \cdot C_{operations} > C_{operations} + C_{extra}$, that is if

$$p > 1 + (C_{extra} / C_{operations}). \quad (4.3)$$

On the other hand, according to (4.1) and (4.2)

$$T_{sequential} = p \cdot T_{clk} > p \cdot (T_{operations} + T_{extra}) > p \cdot T_{operations} \geq T_{iterative}$$
$$(4.4)$$

so that the iterative implementation is faster than the sequential one.

An alternative option consists in partially unroll the *for* loop (De Micheli 1994; Parhami 2000). Assume that $p = k \cdot s$. Then, s successive iteration steps are executed at each clock cycle. An example, with $s = 3$, is shown in Fig. 4.8.

Obviously, the clock cycle, say $T_{clk'}$, must be longer than in the sequential implementation of

Fig. 4.7b (T_{clk}). Nevertheless, it will be generally shorter than $s \cdot T_{clk}$. On the one hand, as already quoted above, if *operations* is a combinational component, the critical path length of s serially connected components is generally shorter than the critical path length of a single component, multiplied by s. On the other hand, the additional register delays are associated with groups of s components so that their impact is divided by s. In other words, $T_{clk'} \cong s \cdot T_{operations} + T_{reg} < s \cdot (T_{operations} + T_{reg}) \cong s \cdot T_{clk}$. Furthermore, when interconnecting several circuits, some additional logical simplifications can be performed by the synthesis tool and have positive repercussions on both the cost and the delay. So

$$C_{partially\_unrolled} \leq s \cdot C_{operations} + C_{extra},$$
$$T_{partially\_unrolled} = (p/s) \cdot T_{clk'}, \quad (4.5)$$

where

$$T_{clk'} < s \cdot T_{clk}. \quad (4.6)$$

Thus, according to (4.1) and (4.5), the partially unrolled implementation has a lower cost than the iterative implementation if $p \cdot C_{operations} > s \cdot C_{operations} + C_{extra}$, that is if

$$p > s + (C_{extra} / C_{operations}). \quad (4.7)$$

On the other hand, according to (4.2), (4.5) and (4.6),

$$T_{partially\_unrolled} = (p/s) \cdot T'_{clk} < (p/s) \cdot s \cdot T_{clk}$$
$$= p \cdot T_{clk} = T_{sequential}.$$
$$(4.8)$$

4.3 Pipelined Implementation of Loops

In Chap. 3, a circuit made up of a set of serially interconnected subcircuits (Fig. 3.6a) has been implemented by a pipelined circuit (Fig. 3.6b). This method can obviously be applied to circuits that implement completely unrolled loops under the form of iterative circuits (e.g., Figs. 4.4 and 4.7a).

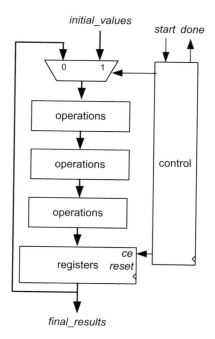

Fig. 4.8 Partially unrolled loop implementation ($s = 3$)

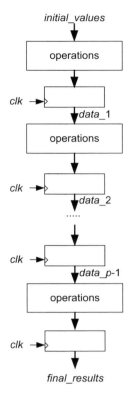

Fig. 4.9 Pipelined circuit

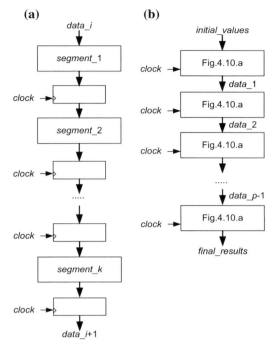

Fig. 4.10 Pipelined version of *operations*

As an example, consider the iterative implementation of Fig. 4.7a. The corresponding pipelined circuit is shown in Fig. 4.9. The data introduction interval is equal to $T_{clk} > T_{operation}$ instead of $p \cdot T_{operation}$ in the case of the circuit of Fig. 4.7a. Furthermore, the power consumption could be reduced because the synchronization barriers (pipeline registers) reduce the generation of spikes.

Another interesting point is the possibility to use, or not, a pipelined version of the component *operations* itself. Consider a pipelined version (Fig. 4.10a) with k segments and a pipeline clock period equal to t_{clock}. Then, the circuit of Fig. 4.7a can be implemented as shown in Fig. 4.10b. The data introduction interval of this circuit is equal to t_{clock}, and its latency is equal to $k \cdot p \cdot t_{clock}$.

Consider now the sequential circuit of Fig. 4.7b. The pipelined version of *operations* might be used, but generally this will not reduce the data introduction interval. The problem is the data dependency between successive executions of the procedure *operations*: at the beginning of iteration number i, the input data of *operations* is $data_{i-1}$; the value of the corresponding output is $data_i$ and is the input data of the next procedure execution; however, $data_i$ will be available only after k clock cycles. Thus, with respect to the data introduction interval, the use of a pipelined component has no effect: the execution of iteration number i cannot start before the execution of iteration number $i - 1$ has been completed.

On the other hand, if there are no data dependencies, then both the data introduction interval and the latency can be reduced. Consider the following generic algorithm in which $data_{in}$ and $data_{out}$ are p-component vectors.

Algorithm 4.5 Generic iterative without data dependency

```
data_in = input_value;
for i in 1 .. p loop
  operations(data_in(i), data_out(i));
end loop;
```

If a non-pipelined component *operations* with computation time equal to $T_{operations}$ is used (Fig. 4.11a), the minimum data introduction interval is $T_{operations}$, and the latency is equal to

$$latency_{non-pipelined} = p \cdot T_{operations}. \quad (4.9)$$

If a k-stage pipelined component *operations* with computation time equal to $k \cdot t_{clk} \cong T_{operations}$ is used (Fig. 4.11b), the minimum data introduction interval is t_{clk} and the latency is computed as follows: a first output $[data_{out}(1)]$ is generated after k clock periods. Then, a new output $[data_{out}(2), data_{out}(3),\dots, data_{out}(p)]$ is generated every clock period. Thus,

$$\begin{aligned} latency_{pipelined} &= (k+p-1) \cdot t_{clk} < p \cdot k \cdot t_{clk} \\ &\cong p \cdot T_{operations} \\ &= latency_{non-pipelined}. \end{aligned}$$
$$(4.10)$$

Example 4.1 Generate a pipelined implementation of Algorithm 4.2 (base-2 logarithm computation). For that the iterative implementation of Fig. 4.4 is modified: parallel registers are inserted between successive computation resources. Furthermore, in order to synchronize the outputs, p deskewing shift registers (Sect. 3.4) are added. The resulting circuit is shown in Fig. 4.12.

(a)

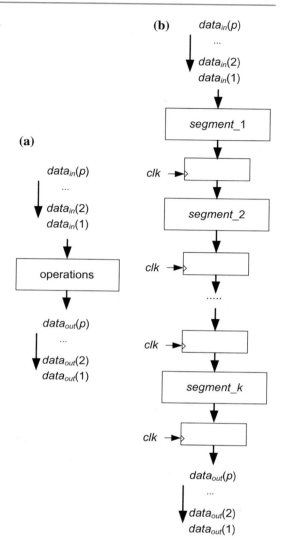

Fig. 4.11 Implementation of Algorithm 4.5

Fig. 4.12 Base-2 logarithm computation: pipelined implementation

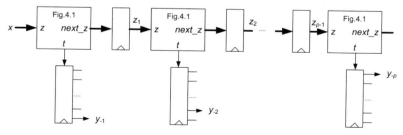

Fig. 4.13 Base-2 logarithm computation: simulation (courtesy of Mentor Graphics)

A VHDL model *logarithm_circuit_pipeline. vhd* has been generated and simulated. It is available at the Authors' web site. The test is executed with the following values: $n = 8$ (number of fractional bits of x), $p = 16$ (number of fractional bits of y) and $m = 16$ (number of fractional bits of z). The circuit repeatedly computes (in hexadecimal)

$$log_2(1.b1) \cong 0.c21a, \ log_2(1.31) \cong 0.40ae,$$
$$log_2(1.f1) \cong 0.f504, log_2(1.05) \cong 0.0724$$

(Fig. 4.13). The number of loop body executions is equal to p so that the latency is equal to $p \cdot T_{clk}$, and the data introduction interval is equal to T_{clk}. The simulation result of Fig. 4.13 shows that once the pipeline is filled ($p = 16$ cycles), a new result is outputted every cycle.

4.4   Digit-Serial Processing

In Sect. 4.2, a partially unrolled version of Algorithm 4.1, namely Algorithm 4.3, has been defined. Comparing both algorithms and both data paths, a conclusion is that in the first case (Algorithm 4.1), one bit t is generated at each step and stored within the output shift register (Fig. 4.2), while in the second case (Algorithm 4.3), two bits t_1 and t_0 are generated at each step and stored within the output shift registers (Fig. 4.6). So, as regards the result generation, the first implementation could be considered as *bit-serial* and the second as *digit-serial*, defining in this case a digit as a 2-bit number.

In fact, the method for computing the base-2 logarithm of a real number presented in Chap. 1 may be modified in such a way that the result is expressed in base 4 instead of 2. Given an n-bit normalized fractional number $x = 1.x_{-1} x_{-2} \cdots x_{-n}$, compute $y = log_2 x$ with an accuracy of k fractional 4-ary digits. As x belongs to the interval $1 \le x < 2$, its base-2 logarithm is a nonnegative number smaller than 1, so $y = 0.y_{-1} y_{-2} \cdots y_{-k}$ where $y_{-i} \in \{0, 1, 2, 3\}$.

If $y = log_2 x$, then $x = 2^{0.y_{-1} y_{-2} \cdots y_{-k} \cdots}$, so that $x^4 = 2^{y_{-1} y_{-2} \cdots y_{-k} \cdots}$. Thus

- if $x^4 \ge 2^3$: $y_{-1} = 3$, $x^4/2^3 = 2^{0.y_{-2} \cdots y_{-k} \cdots}$ and $1 \le x^4/2^3 < 2$;
- if $2^2 \le x^4 < 2^3$: $y_{-1} = 2$, $x^4/2^2 = 2^{0.y_{-2} \cdots y_{-k} \cdots}$ and $1 \le x^4/2^2 < 2$;
- if $2 \le x^4 < 2^2$: $y_{-1} = 1$, $x^4/2 = 2^{0.y_{-2} \cdots y_{-k} \cdots}$ and $1 \le x^4/2 < 2$;
- if $x^4 < 2$: $y_{-1} = 0$ and $x^4 = 2^{0.y_{-2} \cdots y_{-k} \cdots}$.

The following algorithm computes y:

Algorithm 4.6 Base-2 logarithm (digit serial)

```
z = x; i = k;
while i > 0 loop
   if z⁴ ≥ 8 then y_{i-k-1} = 3; z = z⁴/8;
   elsif z⁴ ≥ 4 then y_{i-k-1} = 2; z = z⁴/4;
   elsif z⁴ ≥ 2 then y_{i-k-1} = 1; z = z⁴/2;
   else y_{i-k-1} = 0; z = z⁴;
   end if;
   i = i-1;
end loop;
```

A component that executes the loop body is shown in Fig. 4.14. This component computes a 4-ary output value y_{i-p-1}. The final value $y = 0.y_{-1} y_{-2} \dots y_{-k}$ is expressed in base 4. The translation to a binary number is trivial: every 4-ary digit represents two successive bits of the binary representation. For example, if y is equal to

Fig. 4.14 Digit-serial
computation: computation
resource

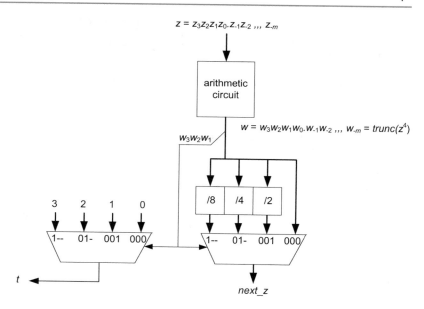

0.20311231 in base 4, then in binary it is equal to 0.10 00 11 01 01 10 11 01. In particular, if y is computed with a precision of k fractional 4-ary digits, the corresponding binary number is computed with a precision of $2 \cdot k$ bits. Thus to obtain the value of $\log_2 x$ with a precision of p fractional bits, the number k of steps is equal to $p/2$.

With this component, several circuit implementations can be considered (iterative, sequential and pipeline). As an example, a pipelined circuit using the component of Fig. 4.14 has been implemented (Fig. 4.15). This circuit is similar to the circuit of Fig. 4.12 with the following differences:

- Every component generates a 4-ary digit (two bits) of the final result so that only $p/2$ components and $p/2-1$ pipeline registers are used.
- The deskewing shift registers store 4-ary digits.

The corresponding VHDL model *logarithm_circuit_pipe_DS.vhd* is available at the Authors' web site. A simulation has been executed with the following values: $n = 8$ (number of fractional bits of x), $p = 16$ (number of fractional bits of y) and $m = 16$ (number of fractional bits of z). The circuit repeatedly computes (in hexadecimal) the same values as in Fig. 4.13. The number of loop body executions is equal to $p/2$ so that the latency is equal to $(p/2) \cdot T_{clk}$, and the data introduction interval is equal to T_{clk}. The simulation result of Fig. 4.16 shows that once the pipeline is filled ($p/2 = 8$ cycles), a new result is outputted every cycle.

The design techniques proposed in this section are commonly used in arithmetic function implementation: an algorithm processes data, or part of them, in a bit-serial manner; a modified version of this initial algorithm permits to process several bits, say D, concurrently. The second

Fig. 4.15 Digit-serial and
pipelined circuit

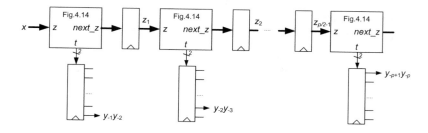

Fig. 4.16 Simulation: pipeline and digit serial (courtesy of Mentor Graphics)

implementation is called *digital-serial,* and D is the digit size.

Loop-unrolling and digit-serial processing are techniques that allow exploring cost–performance tradeoffs, looking for intermediate options between completely combinational (maximum cost and minimum latency) and completely sequential (minimum cost and maximum latency) circuits. Loop unrolling can be directly performed at circuit level, whatever the implemented algorithm, while digit-serial processing looks more like an algorithm transformation. Nevertheless, it is not always so clear that they are different techniques.

4.5 Exercises

1. Given two naturals x and y, with $x < y$, the following *restoring division algorithm* computes two fractional numbers $q = 0.q_{-1} q_{-2} \dots q_{-p}$ and $r < y \cdot 2^{-p}$ such that $x = q \cdot y + r$ and, therefore, $q \le x/y < q + 2^{-p}$:

Algorithm 4.7 Restoring division algorithm

```
r₀ = x;
for i in 1 .. p loop
  z = 2·r_{i-1} - y;
```

```
  if z < 0 then q_{-i} = 0; r_i = 2·r_{i-1};
  else q_{-i} = 1; r_i = z;
  end if;
end loop;
r = r_p·2^{-p};
```

1.1 Define a component that executes the loop body of Algorithm 4.7.
1.2 Implement a circuit that executes Algorithm 4.7.
1.3 Implement unrolled versions of the preceding circuit with different values of s (2, 4, …).
1.4 Define digit-serial versions of Algorithm 4.7 with different values of D (2, 4, …), define components that execute the loop body of the modified algorithms and implement the corresponding circuits.

2. Design other versions (unrolled, digit-serial) of the $log_2 x$ computation circuit.

Bibliography

De Micheli G (1994) Synthesis and Optimization of Digital Circuits. McGraw-Hill, New York.
Parhami B (2000) Computer Arithmetic: Algorithms and Hardware Design. Oxford University Press, New York.

Other Topics of Data Path Synthesis

5

In this chapter, several additional design techniques that permit to optimize some data path features (cost, speed, power consumption), or to make faster and safer the design work, are proposed. The first of them is the use of predefined data path connection structures.

5.1 Data Path Connectivity

Data paths are made of computation resources, registers and connections (Fig. 1.4). This section is dedicated to connections. A generic data path structure is shown in Fig. 5.1. A first connection network permits to transfer register outputs and external inputs to computation resource inputs. A second connection network is used to transfer computation resource outputs to register inputs. This architecture permits to implement sequences of instructions (programs) of the following type:

$$R_i = F(w_0, w_1, \ldots), \ldots, R_j = G(u_0, u_1, \ldots),$$
$$R_k = H(v_0, v_1, \ldots),$$

(5.1)

where $w_0, w_1, \ldots, u_0, u_1, \ldots, v_0, v_1, \ldots \in \{x_0, x_1, \ldots, x_{n-1}, y_0, y_1, \ldots, y_{m-1}\}$, and $F, \ldots, G, H$ are the functions executed by some of the computation resources.

5.1.1 Complete Connectivity

In preceding examples, the connections are implemented by multiplexers. As an example consider the second network of Fig. 5.1:

- For every register R_i, make a list $\{z_j, z_k, \ldots, z_l\}$ of all computation resource outputs that in some cycle of the program execution must be connected to register R_i;
- Then associate with register R_i, a multiplexer with data inputs $z_j, z_k, \ldots, z_l$.

In this way, the second connection network is implemented by m multiplexers each of them with at most k data inputs.

Obviously, the same method can be used to implement the first network. However, in some cases, the computation resource inputs can be interchanged, for example if the corresponding operation is a commutative function. Then an optimization problem must be considered. It can be resolved using graph coloring techniques (as in Figs. 2.16 and 2.19).

In this way, all data transfers necessary to execute instructions such as (5.1) can be performed as soon as the corresponding data (operands or operation results) are available. For that reason,

© Springer Nature Switzerland AG 2019
J.-P. Deschamps et al., *Complex Digital Circuits*,
https://doi.org/10.1007/978-3-030-12653-7_5

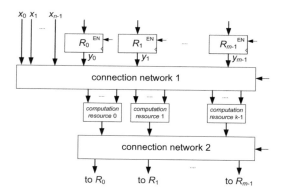

Fig. 5.1 Generic data path structure

this type of connection implementation has been called "complete connectivity."

In order to get an upper bound of the connection network complexity, assume that

- All computation resources have p inputs (or call p, the average number of inputs per computation resource).
- All possible transfers from "register and external inputs" to "resource inputs" and all possible transfers from "resource outputs" to "register inputs" are executed at least once during the program execution, so that to each computation resource input corresponds an $(n + m)$-to-1 multiplexer, and to each register input corresponds a k-to-1 multiplexer.

Then, the number of multiplexer inputs of network 1 is $k \cdot p \cdot (n + m)$ and the number of multiplexer inputs of network 2 is $m \cdot k$. Thus, an upper bound of the number of multiplexer inputs $N_{mux\text{-}inputs}$ is defined by the following condition:

$$N_{mux-inputs} < k \cdot p \cdot (n + m) + m \cdot k. \qquad (5.2)$$

It can easily be proved that an m-to-1 multiplexer can be implemented by $m - 1$ 2-to-1 multiplexers (by induction from $m = 2$). This property justifies that, in what follows, the complexity of a connection network is evaluated by the number of multiplexer inputs.

In fact, a correct cost evaluation should take into account the types of the processed data. The

previous upper bound (5.2) is based on the assumption that all registers, resource inputs and resource outputs have the same number of bits.

Anyway, the upper bound (5.2) is obviously very pessimistic. It corresponds to a circuit that permits to implement $(n + m)^{p \cdot k}$ different connection configurations between $n + m$ signals x_0, x_1, ..., x_{n-1}, y_0, y_1, ..., y_{m-1} and $p \cdot k$ resource inputs, and k^m different connection configurations between k resource outputs z_0, z_1, ..., z_{k-1} and m register inputs. Thus, the total number of connection configurations is the product $(n + m)^{p \cdot k} \cdot k^m$, a huge number generally much greater than the number of different program instructions.

5.1.2 An Optimization Problem

Consider a connection network with r inputs u_0, u_1, ..., u_{r-1} and s outputs v_0, v_1, ..., v_{s-1} (Fig. 5.2) used to implement one of the two connection networks of Fig. 5.1. At each step of the program execution, it must be able to execute in parallel a set of data transfers $\{u_i \rightarrow v_j\}$ from an input to an output. The general method proposed in the preceding section could be summarized as follows:

- For every output v_j, make a list $\{u_i\}$ of inputs such that during the program execution there is at least once a transfer of data from u_i to v_j;
- Associate with output v_j a multiplexer whose data inputs are the elements of the list.

However, given a particular output v_j, all data transfers $\{u_i \rightarrow v_j\}$ are generally not executed during the same program execution cycle.

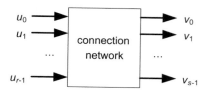

Fig. 5.2 Generic connection network

This suggests a decomposition of the connection network into two sub-circuits.

- First define a partition of the set of inputs $\{u_0, u_1, \ldots, u_{r-1}\}$ into t subsets $S_0, S_1, \ldots, S_{t-1}$, such that if two inputs u_k and u_l belong to the same subset then there is no program execution cycle during which both u_k and u_l are the sources of a data transfer. Then define a first sub-circuit that permits to connect all inputs of a given subset S_l to a common intermediate signal w_l. The sub-circuit is shown in Fig. 5.3a. By definition of the subsets S_l, during every program execution cycle there is at most one element of S_l that is the origin of a data transfer.
- The second sub-circuit must permit to connect the set of intermediate signals $\{w_0, w_1, \ldots, w_{t-1}\}$ to the set of outputs $\{v_0, v_1, \ldots, v_{s-1}\}$. For that, use the same method as before: for every output v_j, make a list $\{w_l\}$ of intermediate signals such that during the program execution there is at least once a transfer of data from one of the inputs of S_l to v_j, and associate with output v_j a multiplexer whose data inputs are the elements of the list (Fig. 5.3b).

Consider a set of data transfer $u_k \rightarrow v_j, \ldots, u_l \rightarrow v_t$ that must be executed during some cycle. Then inputs $u_k, \ldots, u_l$ must belong to different subsets S_l, for example $u_k \in S_p, \ldots, u_l \in S_q$, and the data transfers are executed as follows: $u_k \rightarrow w_p \rightarrow v_j, \ldots, u_l \rightarrow w_q \rightarrow v_t$.

The partition of the set of inputs $\{u_0, u_1, \ldots, u_{r-1}\}$ into t subsets can be stated as a graph coloring problem:

- Define an incompatibility relation over the set of inputs $\{u_0, u_1, \ldots, u_{r-1}\}$: u_k and u_l are incompatible if during some cycle u_k and u_l are the source of a data transfer;
- Color the corresponding graph; assume that there are t different colors $c_0, c_1, \ldots, c_{t-1}$;
- S_l is the set of inputs whose color is c_l.

The number $N_{mux\text{-}inputs}$ of multiplexer inputs of the circuit of Fig. 5.3 satisfies the following condition:

$$N_{mux-inputs} < r + t \cdot s, \qquad (5.3)$$

where r is the number of multiplexer inputs of the circuit of Fig. 5.3a and $t \cdot s$ is an upper bound of the number of multiplexer inputs of the circuit of Fig. 5.3.b. The upper bound (5.3) is smaller than $r \cdot s$, if the number t of subsets S_l is smaller than $r \cdot (s - 1)/s \cong r$.

Example 5.1 Consider the circuit of Fig. 2.21. It connects signals Z (*product*), *adder_out*, *square*, z_A, z_B and R to inputs of registers that store x_A, x_B, z_A, z_B and R. It is the connection network 2 (Fig. 5.1) of a circuit that implements Algorithm 2.4. The inputs of this connection network are $\{Z, adder\_out, sqsuare, z_A, z_B, R\}$, and its outputs are $\{x_A, x_B, z_A, z_B, R\}$. During Algorithm 2.4 execution, the data transfers from inputs to outputs of this connection network, at each step, are the following:

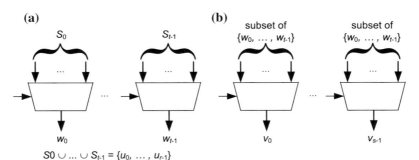

(a)

S_0 S_{t-1}

w_0 w_{t-1}

$S0 \cup \ldots \cup S_{t-1} = \{u_0, \ldots, u_{r-1}\}$

(b)

subset of $\{w_0, \ldots, w_{t-1}\}$ subset of $\{w_0, \ldots, w_{t-1}\}$

v_0 v_{s-1}

Fig. 5.3 Decomposition into two blocks

```
0 to 3: none;
4:    z_B = adder_out;
5:    z_B = square;
6:    z_B = square;
7:    none
8:    R = Z;
9 to 10: none
11:   x_B = Z;
12:   z_A = adder_out;
13:   z_A = square;
14:   none;
15:   x_A = Z;
16:   R = square;
17:   none;
18:   x_B = Z;
19 to 20: none;
21:   x_A = Z;
22:   (x_A, z_A, x_B, z_B) = (z_B, R, ad-
der_out, z_A);
23:   z_A = adder_out;
24:   z_A = square;
25:   z_A = square;
26:   none;
27:   R = Z;
28 to 29: none;
30:   x_A = Z;
31:   z_B = adder_out;
32:   z_B = square;
33:   none:
34:   x_B = Z;
35:   R = square;
36:   none;
37:   x_A = Z;
38 to 39: none;
40:   x_B = Z;
```

```
41:   (x_B, z_B, x_A, z_A) = (z_A, R, ad-
der_out, z_B);
42:   none;
```

The incompatible inputs are (instructions 22 and 41) z_A, z_B, *adder_out* and R. The graph of the corresponding incompatibility relation is shown in Fig. 5.4a. It can be colored with four colors c_0, c_1, c_2 and c_3 to which correspond the following subsets of compatible inputs:

$$S_0 = \{square, adder\_out, Z\}, S_1 = \{z_A\},$$
$$S_2 = \{z_B\}, S_3 = \{R\}.$$

$$(5.4)$$

The circuit of Fig. 5.4b connects the inputs to their corresponding intermediate signal w_0–w_3.

During Algorithm 2.4 execution, the following data transfer must be executed:

$$x_A \leftarrow \{Z, adder\_out, z_B\} \subseteq S_0 \cup S_2,$$
$$x_B \leftarrow \{Z, adder\_out, z_A\} \subseteq S_0 \cup S_1,$$
$$z_A \leftarrow \{adder\_out, square, z_B, R\} \subseteq S_0 \cup S_2 \cup S_3,$$
$$z_B \leftarrow \{adder\_out, square, z_A, R\} \subseteq S_0 \cup S_1 \cup S_3,$$
$$R \leftarrow \{Z, square\} \subseteq S_0.$$

The circuit of Fig. 5.5 connects the intermediate signals w_0–w_3 to the output signals x_A, x_B, z_A, z_B and R.

The complete circuit (Figs. 5.4b and 5.5) consists of five multiplexers with, in total, thirteen multiplexer inputs, while the circuit of Fig. 2.21 has also five multiplexers but with sixteen multiplexer inputs (actually not a great improvement!).

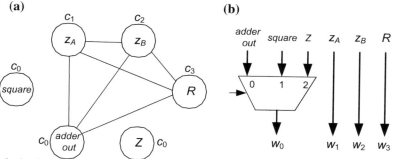

(a)   **(b)**

Fig. 5.4 First sub-circuit

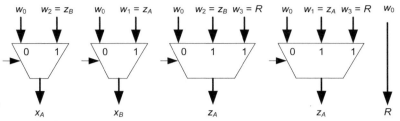

Fig. 5.5 Second sub-circuit

5.1.3 Sequential Implementation

As in many aspects of digital circuit implementation, the reduction in the connection costs is possible if "space" is substituted by "time." A generic example is given in Fig. 5.6. This data path permits to execute the same program as the circuit of Fig. 5.1. It is assumed that all computation resources have p inputs. The connections are executed with an $(n + m)$-to-1 multiplexer, $p - 1$ A-registers with *enable* input, $k - 1$ B-registers with *enable* control input and a k-to-1 multiplexer.

An instruction such as

$$\begin{aligned}
R_i &= F(w_0, \ldots, w_{p-2}, w_{p-1}), \ldots, R_j \\
&= G(u_0, \ldots, u_{p-2}, u_{p-1}), R_l \\
&= H(v_0, \ldots, v_{p-2}, v_{p-1}),
\end{aligned} \tag{5.5}$$

is executed as follows:

- Operands w_0 to w_{p-2} are sequentially transmitted to mux_1 and stored in registers A_0 to A_{p-2};
- Assume that F is the function that corresponds to the computation resource number t;

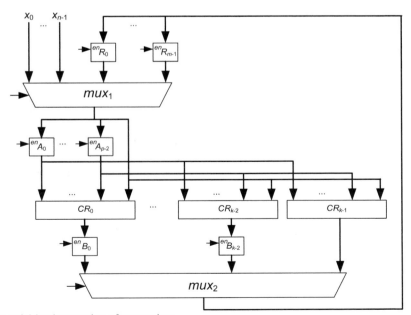

Fig. 5.6 Sequential implementation of connections

the value of the last operand w_{p-1} is transmitted to mux_1 and the value of $F(A_0, A_1, \ldots, A_{p-2}, mux_1) = F(w_0, w_1, \ldots, w_{p-2}, w_{p-1})$ is stored in B_t;

- The values of the next values, up to $G(u_0, \ldots, u_{p-2}, u_{p-1})$, are computed in the same way and are stored in some of the B-registers;
- The value of $H(v_0, \ldots, v_{p-2}, v_{p-1})$ is computed in the same way and directly stored into R_l;
- The other registers' contents $R_i, \ldots, R_j$ are updated with the values stored in B-registers.

In the case where all computation resources are active during an instruction execution, the sequence of operations is the following:

```
A₀ = mux₁ = w₀;
...
A_{p-2} = mux₁ = w_{p-2};
B₀ = CR₀(A₀, ... , A_{p-2}, mux₁) = CR₀(A₀, ... ,
A_{p-2}, w_{p-1});
...
A₀ = mux₁ = u₀;
...
A_{p-2} = mux₁ = u_{p-2};
B_{k-2} = CR_{k-2}(A₀, ... , A_{p-2}, mux₁) = CR_k
-2(A₀, ... , A_{p-2}, u_{p-1});
A₀ = mux₁ = v₀;
...
A_{p-2} = mux₁ = v_{p-2};
R₁ = mux₂ = CR_{k-1}(A₀, ..., A_{p-2}, mux₁) = CR_k
-1(A₀, ..., A_{p-2}, v_{p-1});
R_i = mux₂ = B₀;
...
R_j = mux₂ = B_{k-2};;
```

The total number of steps is

$$N_{steps} = (k-1) \cdot p + (p-1) + m$$
$$= k \cdot p + m - 1 \qquad (5.6)$$

instead of 1 in the case of Fig. 5.1, and there are

$$N_{mux-inputs} = n + m + k \qquad (5.7)$$

multiplexer inputs instead of $k \cdot p \cdot (n + m) + m \cdot k$ in the case of Fig. 5.1. Furthermore, there are $p + k - 2$ additional registers.

Comments 5.1

- Are necessary the B-registers? The problem is "data dependency." Consider an instruction such as (5.5) and assume that u_0 is the value stored in R_i. In the case of the data path of Fig. 5.1, all operations are executed in parallel so that G is computed with the latest (not updated) value of $u_0 = R_i$. In the case of the data path of Fig. 5.6, the operations are sequentially executed. If the value of $F(w_0, \ldots, w_{p-1})$ is directly stored in R_i, then G would be computed with the next (updated) value of $u_0 = R_i$. In some cases, the problem can be avoided by a convenient choice of the order in which the operations are executed. In other cases, the parallel behavior can be emulated by inserting additional intermediate registers (the B-registers of Fig. 5.6).
- As already mentioned in Chapter 3, in the case of FPGA implementations, additional registers does not necessarily increase the total cost, computed in terms of used basic cells. The additional registers could consist of otherwise-not-used flip-flops.

A frequent particular case is when all computation resources have at most two operands and there are no data dependency problems so that the B-registers are no longer necessary. The circuit is shown in Fig. 5.7. The connections are implemented with two multiplexers and a register A (accumulator register).

The sequence of operations is the following:

```
A = mux₁ = w₀;
R_i = mux2 = CR₀(A, mux₁) = CR₀(A, w₁);
...
A = mux₁ = u₀;
R_j = mux2 = CR_{k-2}(A, mux₁) = CR_{k-2}(A, u₁);
A = mux₁ = v₀;
R₁ = mux2 = CR_{k-1}(A, mux₁) = CR_{k-1}(A, v₁);
```

Thus

$$N_{steps} = 2 \cdot k \text{ and } N_{mux-inputs} = n + m + k. \quad (5.8)$$

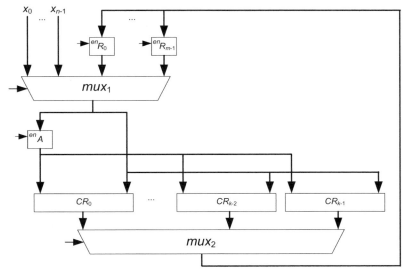

Fig. 5.7 Two-multiplexer and accumulator architecture

A still more particular case is when there is only one computation resource; for example, a programmable arithmetic and logic unit (Sect. 5.3). The circuit is shown in Fig. 5.8. The connections are implemented with a multiplexer and an accumulator register A.

An instruction such as $R_i = F(w_0, w_1)$ is executed in two steps:

```
A = mux₁ = w₀;
Rᵢ = CR(A, mux₁) = CR(A, w₁);
```

and there are $n + m$ multiplexer inputs.

Instead of using multiplexers, a common (traditional, old fashioned) technique is to use buses. As an example, the architecture of Fig. 5.7 is equivalent to the two-bus architecture of Fig. 5.9, with different control signals or with additional address decoders.

Comment 5.2

Within integrated circuits, the internal connections are generally implemented by means of multiplexers instead of busses. A drawback of busses is that, under certain circumstances, busses could be in high impedance, a potentially unstable and undesirable state. This happens if none of the connected 3-state amplifier outputs is in a

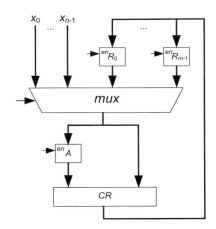

Fig. 5.8 One-multiplexer and accumulator architecture

low-impedance state. To avoid a possible instability, pullup (or pulldown) devices are added.

5.2 Memory Blocks

Part of the registers of a digital system can be grouped together within memory blocks. In this way, the implementation of the corresponding set of registers could be more efficient in terms of silicon area, and the structure of the circuit might be easier to understand and specific control techniques can be considered. All standard cell

Fig. 5.9 Two-bus and accumulator architecture

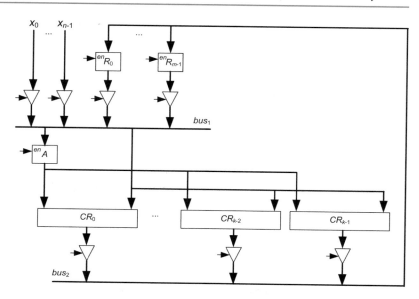

and field programmable gate array libraries include predefined macrocells that implement different types of memory blocks. In this section, some of those memory blocks are described.

5.2.1 Register Files

A register file is a set of m-bit parallel registers plus additional circuits that permit to access the stored data (*read* operation) and to update the register contents (*write* operation). Some of the main parameters of a register file are as follows:

- the number n of registers,
- the number m of bits per register,
- the number of input ports, and
- the number of output ports.

A register file with an input port and an output port is a simple static random access memory that stores 2^n m-bit words. All ASIC and FPGA vendor libraries contain predefined register files that can be instantiated and integrated within the circuit definition. Furthermore, many IC and FPGA developing tools include generators that permit to define customized register files, for example LogiCORE IP Block Memory Generator (Xilinx) and Intel FPGA IP cores (Altera).

Apart from the parameters mentioned here above, there are other features that the designer must specify when defining a register file. Even in the case of single input and output ports, the write and read operations can be controlled in several ways. In the following example (Fig. 5.10), the write operation is synchronous:

Fig. 5.10 Single input and output port register file ($n = 16$)

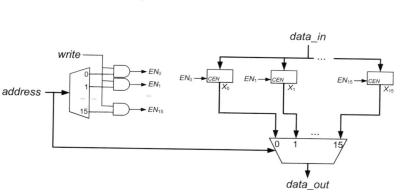

- On a positive edge of *clk,* the value of *data_in* is stored within register number *i* if *address* = *i* and *write* = 1.

On the other hand, the read operation is asynchronous:

- If *address* = *i*, then *data_out* is equal to the value stored in register number *i*.

A VHDL model *register_file.vhd* is available at the Authors' web site.

Example 5.2 The circuit of Fig. 5.11 is a register file with two input ports and two output ports. Inputs and outputs are synchronized. The write operations are executed as follows:

- The enable signal *en_i* of register number *i* is equal to 1, if either *write_A* = 1 and *address_A* = *i* or when *write_B* = 1 and *address_B* = *i*;
- The data inputted to register number *i* is defined by Table 5.1.

In order to avoid a conflict when trying to transfer *input_A* to register number *i* and to transfer *input_B* to the same register and at the same time, it should be assumed that the product *write_A* · (*address_A* = *i*) ·*write_B* · (*address_B* = *i*) is always equal to 0, for all *i*. Anyway, in the case of Fig. 5.11, if *write_A* · (*address_A* = *i*) and *write_B* · (*address_B* = *i*) are equal to 1, then the data stored into register number *i* is *input_B*. So, when generating the data sheet of the preceding circuit, the following rule could be added: simultaneous *write_A* and *write_B* operations to the same register are allowed only if *input_A* = *input_B*.

The read operation uses the same addresses as the *write* operation (*address_A* and *address_B*); it is executed as follows:

- The enable signal of output register *A* is equal to *read_A*, and the enable signal of output register *B* is equal to *read_B*;
- The data stored in output register *A* is selected by a multiplexer controlled by *address_A*,

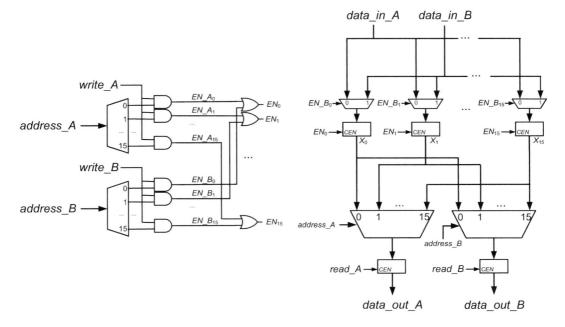

Fig. 5.11 Two-port register file (*n* = 16)

Table 5.1 Input data selection

$write\_A \cdot (address\_A = i)$	$write\_B \cdot (address\_B = i)$	Input to register n° i
0	0	Don't care
0	1	$data\_B$
1	0	$data\_A$
1	1	Not allowed

and the data stored into output register B is selected by another multiplexer controlled by $address\_B$.

A VHDL model *two_port_register_file.vhd* is available at the Authors' web site.

Many different types of register files can be synthesized. As already mentioned, most ASIC and FPGA vendors have development tools that include generators of customized register files. According to the circuit specification, the designer can choose the most convenient parameters, for example: number of registers, number of bits, number of input and output ports, synchronized inputs, registered outputs. As an example, the processor described in Chapter 5 of Deschamps et al. (2017) includes a register file (Fig. 5.16 of Deschamps et al. 2017) with the following characteristics: $n = 16$, $m = 8$, one input port, two output ports, three addresses i (first output port), j (second output port) and k (input port). The output ports are not registered so that instructions such as $R_k = F(R_i, R_j)$, where F is a function executed by an external combinational circuit, can be executed in one clock cycle, even if $k = i$ or $k = j$, if the clock period is longer than the combinational circuit delay.

5.2.2 First-In First-Out Memories

A FIFO is also a set of n m-bit parallel registers but with restrictive access to the stored data. It works as a queue. Assume that it currently stores s data $d_0, d_1, \ldots, d_{s-1}$ with $s < n$. After a *write* operation with input data equal to d_s, the new register contents will be $d_0, d_1, \ldots, d_{s-1}, d_s$. Conversely, starting from the preceding internal state, after a *read* operation the new register contents will be $d_1, d_2, \ldots, d_s$. A graphical description of successive contents of a FIFO is shown in Fig. 5.12.

The external control of a FIFO memory only uses two control signals: *write* and *read*. There are no address bits. On the other hand, two condition signals (flags) are necessary:

- A *full* flag indicates that the memory already stores n data so that a *write* operation is not allowed;
- An *empty* flag indicates that the memory does not store any data so that a *read* operation is not allowed.

A simple way to implement an "address-less" memory is to use a register file and to internally

Fig. 5.12 FIFO file: example of successive states

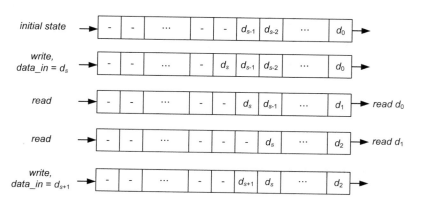

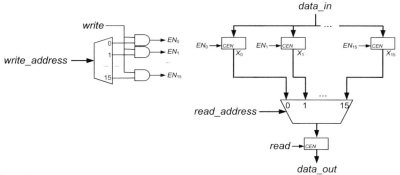

Fig. 5.13 Register file

generate the write and read addresses according to some strategy, in this case a First-In First-Out strategy. As an example, consider a register file with an input port *data_in*, a registered output port *data_out*, two commands *write* and *read*, and two addresses *write_address* and *read_address* (Fig. 5.13).

The complete FIFO structure is shown in Fig. 5.14. The control circuit stores and updates *write_address*, *read_address* and the number s of currently stored data; it generates the *write* and *read* commands as well as the *empty* and *full* flags. In answer to an external *write* command, the following operations are executed:

- The external value *data_in* is stored at address *write_address* of the register file;
- The *empty* flag is set to *false*;
- If the current value of s is $n-1$, the *full* flag is set to *true*;
- *write_address* and s are updated: *write_address* = *write_address* + 1; $s = s + 1$.

In answer to an external *read* command, the following operations are executed:

- The data stored at address *read_address* is sent to the output register.
- The *full* flag is set to *false*.
- If the current value of s is 1 the *empty* flag is set to *true*.
- *read_address* and s are updated: *read_address* = *read_address* + 1; $s = s - 1$.

The behavior of the control circuit is defined by the following algorithm.

Algorithm 5.1 First-In First-Out memory: control unit

```
-- on reset:
s = 0; empty = true; full = false;
write_address = 0; read_address = 0;
loop
   if write = 1 then
     empty = false;
     if s = n-1 then full = true; end if;
     write_address = (write_ad-
dress + 1) mod n;
     s = s + 1;
   elsif read = 1 then
     full = false;
     if s = 1 then empty = true; end if;
     read_address = (read_ad-
dress +1) mod n;
     s = s - 1;
   end if;
end loop;
```

Examples of operations are shown in Fig. 5.15:

- Initially, the queue contents are $d_4\ d_3\ d_2\ d_1$ that are stored between addresses *read_address* and *write_address* − 1; d_1 is the soonest stored data (First-In) and d_4 is the latest stored data; the previously read values are d_0 (the latest) $d_{-1}\ d_{-2}\ d_{-3}$;

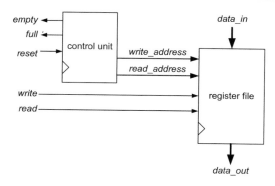

Fig. 5.14   FIFO structure

- After a first *read* operation, the soonest stored data d_1 is outputted and the new queue contents are $d_4\ d_3\ d_2$; the soonest stored data is now d_2;
- After a second *read* operation, the soonest stored data d_2 is outputted and the new queue contents are $d_4\ d_3$;
- After a *write* operation with *data_in* = d_5, the new queue contents are $d_5\ d_4\ d_3$ being d_5 the latest stored data; the latest read value is still d_2.

A VHDL model *fifo.vhd* is available at the Authors' web site. Simulation results are shown in Fig. 5.16:

- Initially   *write_address* = 0,   *read_address* = 0 and $s = 0$.

- Then ten write cycles with *data_in* = 00, 01, 02, …, 09 are executed; at the end of those operations   *write_address* = 10,   *read_address* = 0 and $s = 10$.
- Five read cycles are executed; *data_out* = 00, 01, 02, 03, 04;   *write_address* = 10, *read_address* = 5 and $s = 5$.
- Nine write cycles with *data_in* = 0A, 0B, …, 0F, 10, 11, 12 are executed; *write_address* = (10 + 9) mod 16 = 3, *read_address* = 5 and $s = 5 + 9 = 14$.
- Ten read cycles are executed; *data_out* = 05, 06, …, 0E; *write_address* = 3, *read_address* = 5 + 10 = 15 and $s = 14 - 10 = 4$.

Observe   that $s = (write\_address - read\_address)$ mod n.

An interesting application of FIFO memories (queues) is the implementation of flexible connections between circuits (Fig. 5.17a). Assume that circuits A and B sequentially process data and that their computation times are data-dependent. It can happen that in some moment circuit A has completed a computation and has generated an output data, while circuit B is still processing the previous data. Then circuit A must wait for circuit B being ready before sending it this new data. Conversely in some other moment circuit B could be waiting for circuit A having completed a computation. In such a case, a FIFO (queue) could be inserted

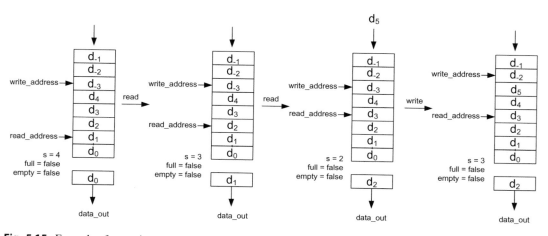

Fig. 5.15   Example of operations

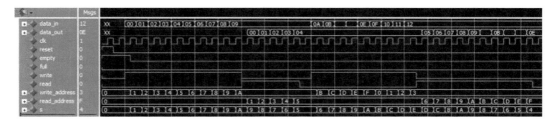

Fig. 5.16 FIFO simulation ($n = 16$, $m = 8$) (courtesy of Mentor Graphics)

Fig. 5.17 Flexible connection

between circuits A and B (Fig. 5.17b): when circuit A generates a new data, it sends it to the queue; when circuit B is ready to process a new data, it reads it from the queue. In this way, as long as the queue is neither full nor empty, none of circuits A and B will have to wait before starting a new computation.

This type of interconnection can be used in pipelined systems (flexible pipelining) or in input–output interfaces.

5.2.3 First-In Last-Out Memories

A FILO (or LIFO = Last-In First-Out) is also a set of n m-bit parallel registers with restrictive access to the stored data. It works as a stack. Assume that it currently stores s data d_0, d_1, ..., d_{s-1} with $s < n$. After a *write* operation with input data equal to d_s, the new register contents will be d_0, d_1, ..., d_{s-1}, d_s. Conversely, starting from the preceding internal state, after a *read* operation the new register contents will be d_0, d_1, ..., d_{s-1}. A graphical description of successive contents of a FILO is shown in Fig. 5.18.

The external control of a FILO memory uses two control signals *write* (or *push*) and *read* (or *pop*) and generates two condition signals (flags) *full* and *empty*.

As in the case of a FIFO memory, a simple way to implement a FILO memory is to use a

register file and to internally generate the write and read addresses. As an example, consider again the register file of Fig. 5.13 and the structure of Fig. 5.19. The *write_address* points to the register where a new data must be written, while the *read_address* points to the register where the latest data has been written. In answer to an external *write* command, the following operations are executed:

- The external value *data_in* is stored at address *write_address* of the register file.
- The *empty* flag is set to *false*.
- If the current value of *write_address* is n 1, then the *full* flag is set to *true*.
- *write_address* is updated: *write_address* = *write_address* + 1.

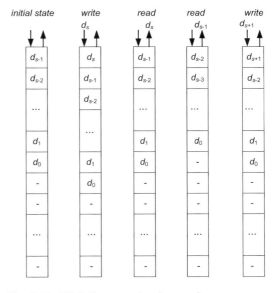

Fig. 5.18 FILO file: example of successive states

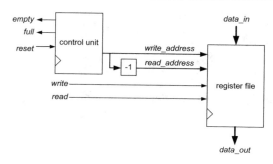

Fig. 5.19 FILO structure

In answer to an external *read* command, the following operations are executed:

- The data stored at address *read_address* = *write_address* − 1 is sent to the output register.
- The *full* flag is set to *false*;
- If the current value of *write_address* is 1, then the *empty* flag is set to *true*;
- *write_address* is updated: *write_address* = *write_address* − 1.

The behavior of the control circuit is defined by the following algorithm.

Algorithm 5.2 First-In Last-Out memory: control unit

```
-- on reset:
empty = true; full = false; write_ad-
dress = 0;
loop
  if write = 1 then
    empty = false;
    if write_address = n-1 then full =
true; end if;
    write_address = (write_address + 1)
mod n;
  elsif read = 1 then
    full = false;
    if write_ad-
dress = 1 then empty = true; end if;
    write_address = write_address -1;
  end if;
end loop;
```

Examples of operations are shown in Fig. 5.20:

- Initially, the stack contents are $d_4\, d_3\, d_2\, d_1\, d_0$ that are stored between addresses 0 and *read_address*; d_4 is the latest stored data (Last-In) and d_0 is the soonest stored data.
- After a *write* operation with *data_in* = d_5, the new queue contents are $d_5\, d_4\, d_3\, d_2\, d_1\, d_0$ and the latest stored data is now d_5.
- After a *read* operation the latest stored data d_5 is outputted and the new queue contents are $d_4\, d_3\, d_2\, d_1\, d_0$.
- After another *read* operation, the latest stored data d_4 is outputted and the new queue contents are $d_3\, d_2\, d_1\, d_0$.

A VHDL model *filo.vhd* is available at the Authors' web site. Simulation results are shown in Fig. 5.21:

- initially *write_address* = 0;
- Then ten write cycles with *data_in* = 00, 01, 02, …, 09 are executed; at the end of those operations *write_address* = 10 and *read_address* = 9.
- Five read cycles are executed; *data_out* = 09, 08, 07, 06, 05; *write_address* = 10 − 5 = 5 and *read_address* = 4.
- Nine write cycles with *data_in* = 0A, 0B, …, 0F, 10, 11, 12 are executed; *write_address* = 5 + 9 = 14 and *read_address* = 13.
- Ten read cycles are executed; *data_out* = 12, 11, …, 0A, 04; *write_address* = 14 − 10 = 4 and *read_address* = 3.

A common application of stacks is the storing of the return addresses in processors whose instruction set includes subroutine calls. In order to permit nested subroutine calls, every time that a call is executed the return address is pushed onto the stack. When executing the corresponding "end of subroutine" instruction, the next address is read from the stack. Furthermore, not only the return addresses but also other context data, for example current value of the state registers, can be saved onto the stack when executing a call

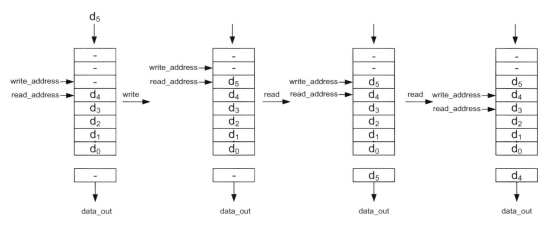

Fig. 5.20 Example of operations

and retrieved from the stack when completing the subroutine execution.

Another (somewhat exotic) application is the evaluation of arithmetic formulas expressed in polish notation. Consider a string of numbers and operands such as

$$+ d/c + a\ b. \qquad (5.9)$$

In this type of expression, the binary operators (+ and / in this example) precede the corresponding operands. For example

$$+a\ b = (a+b);$$
$$/c + a\ b = /c(a+b) = c/(a+b);$$
$$+d/c + a\ b = +d(c/(a+b)) = d + (c/(a+b)). \qquad (5.10)$$

This computation can be executed with a stack:

- Read the characters (operators and operands) from right to left.
- If the read character is an operand, push it onto the stack.
- If the read character is a binary operator, read the two operands that are on the top of the stack, execute the operation with those operands, and push the result onto the stack.
- After processing the leftmost character, the result is in top of the stack.

The successive stack contents when computing expression (5.9) are shown in Fig. 5.22.

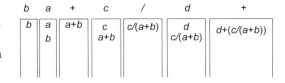

Fig. 5.22 $+ d/c + a\ b = d + (c/(a + b))$

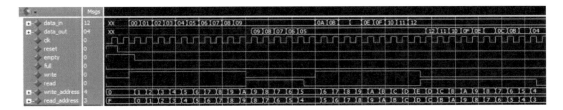

Fig. 5.21   FILO simulation ($n = 16$, $m = 8$) (courtesy of Mentor Graphics)

5.3 Programmable Computation Resources

Data paths (see, for example, the generic data path of Fig. 5.1) include computation resources that implement the computation primitives (operations) corresponding to the executed instructions. The number and type of computation resources mainly depend on the chosen schedule and on the resource assignment (Chap. 2). A key concept is that of *activity intervals* of every computation primitive f (Sect. 2.3).

Consider now a computation resource CR_i that implements a primitive f. The activity intervals of CR_i can also be defined: they consist of all activity intervals of f to which this particular resource CR_i has been assigned.

Example 5.3 Consider the scheduled precedence graph of Fig. 2.8. The activity intervals of the multiplication are $[1, M]$, $[1, M]$, $[1, M]$, $[M + 3, 2M + 2]$, $[M + 1, 2M]$. Three multiplications must be executed during cycles 1 to M, so that three computation resources (multipliers) CR_1, CR_2 and CR_3 are necessary to execute the program. Their activity intervals can be chosen as follows: CR_1: $[1, M]$, $[M + 3, 2M + 2]$, CR_2: $[1, M]$, $[M + 1, 2M]$, CR_3: $[1, M]$.

Consider a particular data path. It can happen that two or more computation resources, say CR_i, CR_j, …, CR_k, implement different functions, say $f_i, f_j, …, f_k$, but that their activity intervals do not overlap: during cycles when CR_i computes f_i,

none of CR_j, …, CR_k is active; during cycles when CR_j executes f_j, none of CR_i, …, CR_k is active, and so on. Then instead of implementing $f_i, f_j, …, f_k$ with separate resources CR_i, CR_j, …, CR_k, an alternative solution can be considered: define a programmable resource PCR able to execute either f_i or f_j or … or f_k under the control of signals generated by the control unit.

Example 5.4 Design a circuit that computes $z = (a + b) - (c + d)$. Assume that each operation ($+$ or $-$) is executed in a cycle and that a 2-cycle schedule has been chosen (actually the ASAP schedule):

```
cycle0: e = a + b;   f = c + d;
cycle1: z = e - f;
```

The computation width with respect to the addition is equal to 2, and the computation width with respect to the difference is equal to 1. The corresponding circuit must include two adders and a subtractor (Fig. 5.23a). Taking into account that the adders are active during cycle 0 and the subtractor is active during cycle 1, a programmable adder–subtractor could be considered: it computes $e = a + b$ during cycle 0 and computes $z = e - f$ during cycle 1. The corresponding circuit is shown in Fig. 5.23b. The programmable resource is an adder–subtractor whose function (*add* or *subtract*) is defined by a control signal *cycle* generated by the control unit.

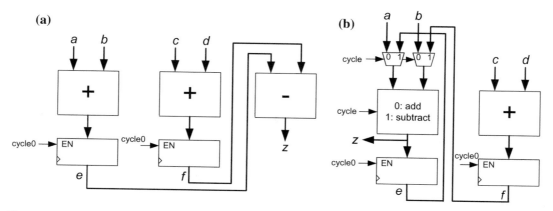

Fig. 5.23 $z = (a + b) - (c + d)$

The definition of programmable resources may have positive effects. In some cases, the cost of a programmable resource is lower than the sum of costs of the non-programmable substituted resources. This often happens in the case of arithmetic circuits when a common kernel can be extracted from a set of arithmetic functions, or when one of the functions is a primitive used to compute another function.

A simple and classical example is the adder–subtractor. Assume that x and y are n-bit natural (non-negative) numbers. Their sum is an $(n + 1)$-bit natural and their difference is a 2's complement $(n + 1)$-bit integer:

$$\text{if } 0 \leq x, y < 2^{n-1} \text{ then } 0 \leq x + y < 2^n \text{ and}$$
$$-2^{n-1} < x - y < 2^{n-1}. \tag{5.11}$$

Their difference is computed as follows:

$$x - y = x + \bar{y} + 1 \tag{5.12}$$

where $\bar{y}$ stands for the bitwise complement of y. It is easy to check that $\bar{y} = 2^n - 1 - y$. The structure of an adder–subtractor is shown in Fig. 5.24. Its cost is practically equal to the cost of a simple adder. It works as follows:

- If $oper = 0$ then z_{n-1} z_{n-2} $\ldots$ $z_0 = (x + y) \bmod 2^n$ and $c_{out} \oplus oper = c_{out}$ is equal to 1 if $x + y \geq 2^n$, so that $c_{out} \oplus oper$ is the most significant bit z_n of $x + y$.

- If $oper = 1$ then z_{n-1} z_{n-2} $\ldots$ $z_0 = (x + (2^n - 1 - y) + 1) \bmod 2^n = (x - y) \bmod 2^n$ and $c_{out} \oplus oper = not(c_{out})$ is equal to 1 if $c_{out} = 0$, that is if $x + (2^n - 1 - y) + 1 < 2^n$ and thus $x - y < 0$, so that z_n is the sign bit of $x - y$.

A straightforward generalization of the adder–subtractor is the arithmetic and logic unit (ALU). It is a basic component of any processor. To complete this section, an ALU kernel is designed.

The circuit to be developed (Fig. 5.25) is a combinational one. Its inputs and outputs are the following:

- two m-bit data inputs x and y;
- an m-bit data output z;
- two input control signals: $oper$ (1 bit) and $select$ (2 bits);
- four output state signals: $carry$, $overflow$, $negative$ and $zero$.

Its behavior is defined in Table 5.2 in which and, or and $not(^-)$ are bitwise logic operations.

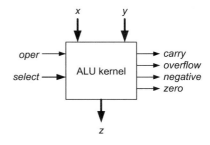

Fig. 5.25 Kernel of an arithmetic and logic unit

Table 5.2 ALU kernel operations

oper	select	Operation
0	00	$z = (x + y) \bmod 2^m$
0	01	$z = x \text{ and } y$
0	10	$z = x \text{ or } y$
1	00	$z = (x - y) \bmod 2^m$
1	01	$z = x \text{ and } \bar{y}$
1	10	$z = x \text{ or } \bar{y}$

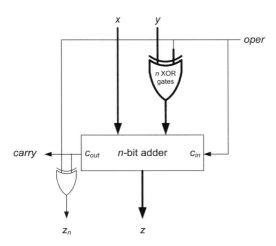

Fig. 5.24 Adder–subtractor

The state output signals (flags) are defined as follows:

- *carry*: if x and y are interpreted as m-bit naturals, *carry* = 1 if $x + y \geq 2^m$ (*oper* = 0) or if $x + y \geq 0$ (*oper* = 1);
- *overflow*: if x and y are interpreted as m-bit 2's complement integers, *overflow* = 1 if $x + y$ cannot be represented with only m bits (*oper* = 0) or if $x - y$ cannot be represented with only m bits (*oper* = 1);
- *negative*: if z is interpreted as an m-bit 2's complement integer, *negative* = 1 if $z < 0$;
- *zero*: *zero* = 1 if $z = 00 \cdots 0$.

The circuit structure is shown in Fig. 5.26b. The basic cell (Fig. 5.26a) computes

$$g_i = x_i \cdot y_i \text{(carry generate)},$$
$$p_i = x_i \text{ or } y_i \text{(carry propagate)},$$
$$s_i = x_i \oplus y_i \oplus c_i,$$
$$c_{i+1} = g_i \text{ or } c_i \cdot p_i.$$

Observe that in this case the carry propagate function is defined as being the Boolean sum of x_i and y_i instead of their mod 2 sum.

Assume that x and y are interpreted as m-bit naturals. If *oper* = 0, then the set of basic cells computes $x + y = c_m \cdot 2^m + z$, so that $z = (x + y) \bmod 2^m$ and $c_m = 1$ if $x + y \geq 2^m$. If *oper* = 1, then the set of basic cells computes $x + (2^m - y - 1) + 1 = c_m \cdot 2^m + z$ so that $z = (x - y) \bmod 2^m$ and $c_m = 1$ if $x + (2^m -$

$y - 1) + 1 \geq 2^m$ that is if $x - y \geq 0$. Thus the *carry* flag is equal to c_m.

To define the overflow flag, x and y must be interpreted as m-bit 2'complement integers. Their sum and their difference are $(m + 1)$-bit 2'complement integers. In order to detect an overflow condition, consider the circuit of Fig. 5.27 in which x and y are represented with an additional bit (with $m + 1$ bits). For that, the sign bits x_{m-1} and y_{m-1} are duplicated. If s_m and s_{m-1} are not equal, the result cannot be represented with m bits. Condition $s_m \neq s_{m-1}$ is equivalent to $c_m \neq c_{m-1}$ so that the *overflow* flag is equal to $c_m \oplus c_{m-1}$.

The two remaining flags are *negative* = z_{m-1} (sign bit of z) and *zero* = $nor(z_{m-1}, z_{m-2}, \ldots, z_0)$.

A VHDL model *alu_kernel.vhd* is available at the Authors' web site.

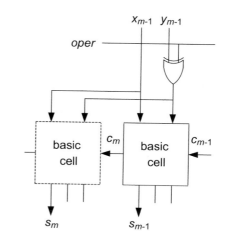

Fig. 5.27 Overflow detection

Fig. 5.26 Arithmetic and logic unit: structural description

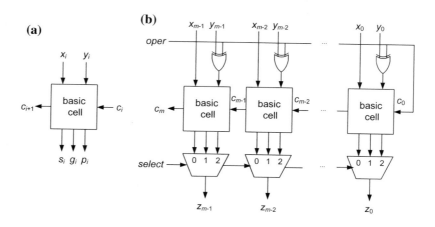

5.4 Sequential Implementation

A central aspect of the development of a digital system is the scheduling of the operations (Chap. 2). The designer looks for a satisfactory trade-off between characteristics such as cost, speed, power consumption, reliability, time to market, and others aspects. If speed is essential, concurrent options based on ASAP schedules should be considered, whatever the cost or the power consumption (obviously within some limits). On the contrary, if the main requirement is to minimize the cost—the silicon area of an IC, the number of FPGA cells—schedules that minimize the computation widths should be chosen.

On the other hand, techniques that permit to reduce the cost of the data path have been described in this chapter: sequential implementation of connections, register files and programmable computation resources. Thus, if cost is the main issue, optimized schedules and sequential implementation techniques might be considered.

As an example, consider a PID (Proportional Integral Derivative) controller loop (Fig. 5.28) used to control a physical (e.g., industrial) process. Its function is to maintain some process parameter $s(t)$—a pressure, a temperature, etc.—as close as possible to a reference value $r(t)$. For that, it periodically calculates an error $e(t)$ equal to the difference $r(t) - s(t)$ and generates a correction $u(t)$ whose effect on the process must be a reduction of the error.

In PID, controllers the correction is a linear combination (K_p, K_i and K_d are constant coefficients)

$$u(t) = K_p e(t) + K_i \int_0^t e(k)dk + K_d \frac{de(t)}{dt}.$$

(5.13)

The integral term makes smoother the transitions between process stable states and the derivative term serves to anticipate the process reactions. Thus, the control is more stable and faster than a simple proportional control.

A discrete version of (5.13) with sampling period T is

$$u(nT) = K_p e(nT)$$
$$+ K_i \sum_{k=0}^{k=n} e(kT)T$$
$$+ K_d \frac{e(nT) - e((n-1)T)}{T}.$$

(5.14)

From Eq. (5.14),

$$u((n-1)T) = K_p e((n-1)T) + K_i \sum_{k=0}^{k=n-1} e(kT)T$$
$$+ K_d \frac{e((n-1)T) - e((n-2)T)}{T}$$

(5.15)

so that (Eqs. 5.14 and 5.15)

$$u(nT) = u((n-1)T) + K_1 \cdot e(nT) + K_2$$
$$\cdot e((n-1)T) + K_3 \cdot e((n-2)T),$$

(5.16)

where

$$K_1 = K_p + K_i \cdot T + K_d/T, K_2$$
$$= -K_p - 2K_d/T, K_3 = K_d/T.$$

(5.17)

A straightforward implementation of (5.16) is shown in Fig. 5.29. For every T seconds, the internal values e_d, e_{dd} and u_d are updated, and new values of r and s are sampled. Then $u = u_d + K_1 \cdot (r - s) + K_2 \cdot e_d + K_3 \cdot e_{dd}$ is computed. An upper bound of the computation time

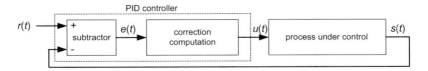

Fig. 5.28 PID controller

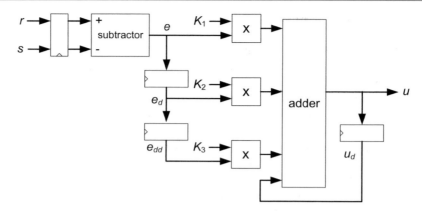

Fig. 5.29 Straightforward implementation

is the sum of three delays: subtractor, multiplier by a constant and 4-operand adder delays.

Practically in all the cases, the sample period T is much longer than the computation time so that sequential implementations could be considered. As an example, the following algorithm executes the PID controller function. It is assumed that an external timer generates a *time_out* pulse every T seconds.

Algorithm 5.3 PID controller

```
loop
  e = r − s;
  acc = ud + K1·e;
  acc = acc + K2·ed;
  u = acc + K3·edd, ud = acc + K3·edd;
  edd = ed;
  ed = e;
  wait until time_out = 1;
end loop;
```

A data path able to execute Algorithm 5.3 is shown in Fig. 5.30. Is consists of

- a programmable computation resource that computes $z = x - y$, $z = x + k_1 \cdot y$, $z = x + k_2 \cdot y$, $z = x + k_3 \cdot y$ and $z = x$, under the control of an *oper* signal (Table 5.3);
- an eight-word register file, with an input port and two output ports that stores the algorithm variables e, e_d, e_{dd}, u_d and *acc* at addresses 0,

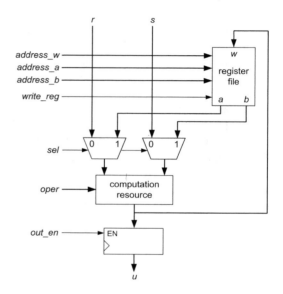

Fig. 5.30 Sequential implementation: data path

Table 5.3 Computation resource operations

oper	z
000	x
010	$x + k_1 \cdot y$
100	$x + k_2 \cdot y$
110	$x + k_3 \cdot y$
001	$x - y$

1, 2, 3 and 7; the outputs are not registered so that operations such as $R_i = f(R_j, R_k)$, where R_i, R_j and R_k are registers of the register file, can be executed in one cycle;

- a connection network consisting of two 2-to-1 multiplexers;
- a control unit that generates *address_w* (input port address), *address_a* and *address_b* (output port addresses), *write_reg*, *sel*, *oper* and *out_en*.

It is assumed that on an initial *reset* pulse the register file contents and the timer are reset.

The control unit has seven internal states that correspond to the seven instructions of the loop body (Fig. 5.31).

Table 5.4 defines the control unit outputs.

A VHDL model *pid_controller.vhd* is available at the Authors' web site. All processed data

Table 5.5 First computation steps

r	s	u_d	e	e_d	e_{dd}	u
7	15	0	−8	0	0	−856
7	5	−856	2	−8	0	190
7	8	190	−1	2	−8	−141
7	7	−141	0	−1	−2	−33
7	7	−33	0	0	−1	−35

(u, u_d, e, e_d, e_{dd}) are represented as m-bit 2's complement integers. This number must be chosen in such a way that there is no overflow. Simulation results are shown in Fig. 5.32. They correspond to the following PID controller:

$$K_1 = 107, K_2 = -104, K_3 = 2, \text{so that u}$$
$$= u_d + 107 \cdot (r - s) - 104 \cdot e_d + 2 \cdot e_{dd}.$$

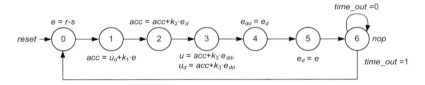

Fig. 5.31 Control unit

Table 5.4 Control signals

Operation	write_reg	sel	address_w	address_a	address_b	oper	en_out
$e = r - s$	1	0	0	−	−	001	0
$acc = u_d + K_1 \cdot e$	1	1	4	3	0	010	0
$acc = acc + K_2 \cdot e_d$	1	1	4	4	1	100	0
$u = u_d = acc + K_3 \cdot e_{dd}$	1	1	3	4	2	110	1
$e_{dd} = e_d$	1	1	2	1	−	000	0
$e_d = e$	1	1	1	0	−	000	0
wait (nop)	0	−	−	−	−	−	0

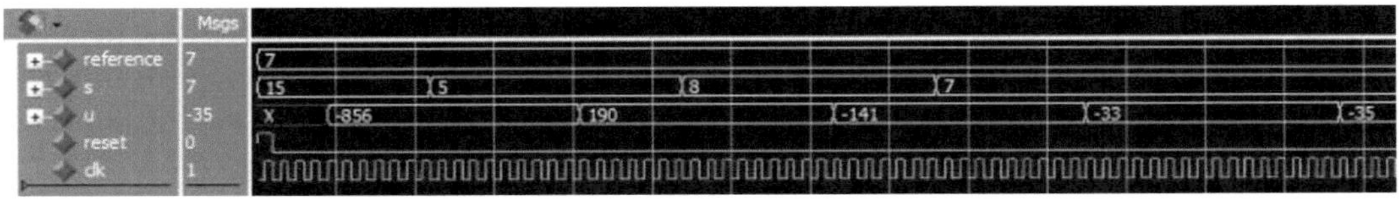

Fig. 5.32 Simulation results (courtesy of Mentor Graphics)

The first computation steps are shown in Table 5.5. Initially, e_d, e_{dd} and u_d are equal to 0 and r is equal to 7. The successive sampled values of s are 15, 5, 8, 7 and 7.

5.5   Hierarchical Description

Hierarchical description and development constitute an efficient strategy in many technical disciplines. In particular, it is commonly used in software engineering as well as in digital system synthesis.

Generally, the initial specification of a digital system is functional (a description of what the system does). In the case of very simple systems, it could be a table that defines the output signal values in function of the input signal values. However, for more complex systems, other specification methods should be used. A natural language description (e.g., in English) is a frequent option. Nevertheless, an algorithmic description (programing language, hardware description language, pseudocode) could be a better choice: those languages have a more precise and unambiguous semantics than natural languages. Furthermore, every block is treated as a subsystem to which a more detailed block diagram is associated, and so on. The design work ends when all block diagrams are made up of interconnected components defined by their function and belonging to some available library of physical components (logic gates, registers, multiplexers, memory blocks, multipliers, dividers, and other cells and macro-cells).

Consider the example of Sect. 5.4 whose initial specification is Algorithm 5.3. From this algorithm, a first data path block diagram (Fig. 5.30) is generated. It includes five blocks: a register file, two 2-to-1 multiplexers, a programmable computation resource and an output register. The output register, the register file and the two multiplexers can be considered as library components (real or virtual compiled components). On the other hand, the programmable computation resource is a functional block whose behavior is defined by Table 5.3. The part of the VHDL code (*pid_controller.vhd*) that corresponds to the programmable resource is the following (*long_x* and *long_z* are 2*m*-bit extensions of *m*-bit integers x and z, necessary for syntax correction):

```
with oper select long_z <=  long_x - one*y when "001",
                            long_x + k1*y when "010",
                            long_x + k2*y when "100",
                            long_x + k3*y when "110",
                            long_x when others;
z <= long_z(m-1 downto 0);
```

programing language and hardware description language specifications can be compiled and executed, so that the initial specification can be tested.

The digital system designer work is the generation of a circuit made up of available components and whose behavior corresponds to the initial specification. Many times this work consists of successive refinements of an initial description: starting from an initial specification, a (top level) block diagram is generated; then,

A straightforward translation of this piece of code to a block diagram is shown in Fig. 5.33. It consists of a 2's complement multiplier that computes $z = x + w \cdot y$ and a multiplexer that selects the value of w.

The 5-to-1 multiplexing function (selection of w) can be implemented by an 8-to-1 multiplexer, that is a library component. It remains to implement an m-bit by m-bit 2's complement multiplier. For that, a modified shift and add algorithm

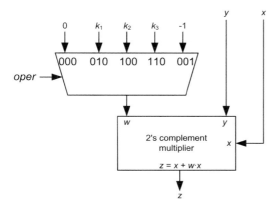

Fig. 5.33 Programmable computation resource

could be used (see, e.g., Sect. 8.4 of Deschamps et al. 2012). The corresponding circuit is an array of $(m + 1)$ by $(m + 1)$ 1-bit multipliers that computes $z = x \cdot y + u + v$ where x, y and u are m-bit 2's complement integers and v is an $(m - 1)$-bit natural (in other words $v_{m-1} = 0$). An

example with $m = 3$ is shown in Fig. 5.34 and can easily be generalized. There are two types of cells: normal 1-bit multiplier cells that compute two switching functions e and f of four binary variables a, b, c and d

$$e = a \cdot b \oplus c \oplus d, f = a \cdot b \cdot c \oplus a \cdot b \cdot d \oplus c \cdot d,$$
(5.18)

and modified 1-bit multiplier cells (last row) that compute

$$e = \bar{a} \cdot b \oplus c \oplus d, f = \bar{a} \cdot b \cdot c \oplus \bar{a} \cdot b \cdot d \oplus c \cdot d.$$
(5.19)

To summarize:

- A block diagram (Fig. 5.30) has been generated from Algorithm 5.3; it is a first hierarchical level; all blocks, but one, can be implemented by real or virtual available components.

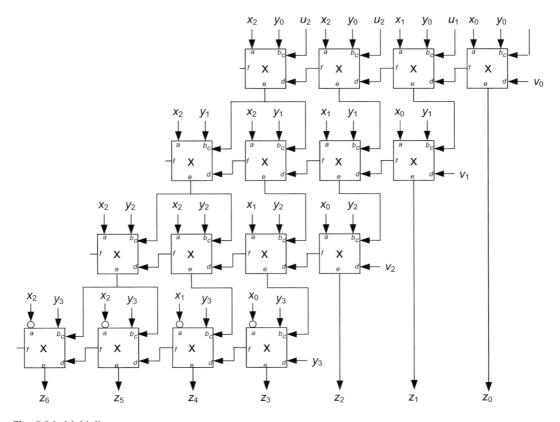

Fig. 5.34 Multiplier array

- The programmable resource behavior is defined by a table (Table 5.3); to this behavior corresponds a block diagram (Fig. 5.33); it is a second hierarchical level.
- At this point, the decision taken by the designer is to use a multiplier that implements the modified shift and add algorithm; the corresponding block diagram is shown in Fig. 5.34; it is an $m + 1$ by $m + 1$ array of simple 1-bit multiplier cells; this is a third hierarchical level;
- There are two types of multiplier cells, and each of them is defined by very simple switching functions (5.18 and 5.19) that can easily be implemented with logic gates or FPGA basic cells; it is a fourth hierarchical level (gate level).

A parameterized VHDL model *modified_parallel_mutiplier.vhd* is available at the Authors' web site. A VHDL model *pid_controller2.vhd* including the programmable resource of Fig. 5.33 and, in particular, the multiplier array of Fig. 5.34 is available at the Authors' web site.

Comment 5.3

When using a hardware description language such as VHDL, hierarchical descriptions are based on the instantiation of components. In the case of the preceding example, a 2's complement multiplier is defined and encapsulated within an entity called *modified_parallel_multiplier*. Then, a *computation_resource* entity is defined; its architecture includes the instantiation of the *modified_parallel_multiplier* component. Finally, a *pid_controller2* entity is defined; its architecture includes the instantiation of the *computation_resource* component (the definition of those entities is available at the Authors' web site).

This method is similar to the use of functions in programming languages. It has the same advantages as in the case of software engineering: clearer description and documentation, faster and safer development task (an application to the engineering world of the Latin "divide et impera" sentence), reuse possibility, and others.

However, sometimes it can be useful to transform the hierarchical description into a flat description. It is an operation similar to *inlining* in the case of software. It (roughly) consists of replacing instantiations of components by the code that describes the component. The reason is that a flat description could give to the synthesis programs more possibilities of low-level optimizations. A very simple example: some pruning operations (elimination of unnecessary gates or other basic cells) are possible if completely flat descriptions are considered. Thus

- Good design practices rely on hierarchical descriptions.
- Subsequent flattening operations, executed by EDA (Electronic Design Automation) tools, can be considered.

5.6 Exercises

1. Define a "one-bus and accumulator" architecture that permits to execute in two clock cycles operations such as $R_i = F(w_0, w_1)$ being R_i a register, w_0 and w_1 either register contents or inputs, and F one of the functions that a programmable computing resource CR can execute.

2. Design a content-addressable memory with an n-bit input *address*, an m-bit input *data* and an n-bit output *&data*. It generates two 1-bit outputs (flags) *match* and *no_match*. It internally stores 2^n m-bit words within a register file R. In *write* mode, it behaves as a conventional register file: $R(address) = data$.

In *read* mode, it compares *data* with the register file contents, from address 0 to address $2^n - 1$. If there is a match at address i, then $\&data = i$ and *match* = 1. If there is no match, *no_match* = 1.

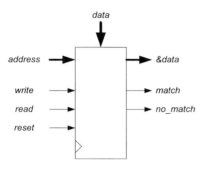

Bibliography

Deschamps JP, Sutter G, Cantó E (2012) Guide to FPGA Implementation of Arithmetic Functions. Springer, Dordrecht

Deschamps JP, Valderrama E, Terés Ll (2017) Digital Systems: from Logic Gates to Processors. Springer, New York

Control Unit Synthesis

Modern electronic design automation (EDA) tools have the capacity to synthesize the control unit from a finite-state machine description or even to extract and synthesize the control unit from a functional description of the complete circuit (Chap. 8). Nevertheless, in some cases, the digital circuit designer can be interested in performing part of the control unit synthesis. Two specific synthesis techniques are presented in this chapter: command encoding and hierarchical decomposition (De Micheli 1994). Both of them pursue a double objective. On the one hand, they aim at reducing the circuit cost. On the other hand, they can make the circuit structure easier to understand and to debug. The latter is probably the most important aspect.

The use of components whose latency is data-dependent has been implicitly dealt with in Sect. 2.5. Some additional comments about variable-latency operations are made in the third section of this chapter.

Cost is generally not an issue in the case of control unit synthesis. However, classical software techniques such as separation of operations and jumps, multi-way branching, subroutines and others could be used to reduce the control unit cost (if necessary).

6.1 Command Encoding

Consider the control unit of Fig. 1.4 and assume that *commands* is an m-bit vector, *conditions* a p-bit vector and *internal_state* an n-bit vector. Thus, the command generation block generates $m + 1$ binary function of $p + n + 1$ binary variables (Fig. 6.1a). Nevertheless, the number s of different commands is generally much smaller than 2^m. An alternative option is to encode the s commands with a t-bit vector, with $2^t \geq s$. The command generation block of Fig. 6.1a can be decomposed into two blocks as shown in Fig. 6.1b: the first one generates $t + 1$ binary functions of $p + n + 1$ variables, and the second one (the command decoder) m binary functions of t binary variables.

A generic circuit complexity measure is the number of bits that a memory (ROM) must store in order to implement the same functions. Thus, the complexity of a circuit implementing a function of $p + n + 1$ variables is 2^{p+n+1} bits (size of the corresponding truth vector), and the complexity of a circuit that implements $m + 1$ function of $p + n + 1$ variables is

$$(m + 1) \cdot 2^{p + n + 1} \text{ bits.} \qquad (6.1)$$

J.-P. Deschamps et al., *Complex Digital Circuits*,
https://doi.org/10.1007/978-3-030-12653-7_6

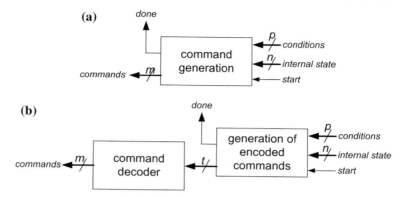

Fig. 6.1 Command generation

The total complexity of two circuits implementing $t + 1$ function of $p + n + 1$ variables and m functions of t variables, respectively, is

$$(t + 1) \cdot 2^{p+n+1} + m \cdot 2^t \text{ bits.} \qquad (6.2)$$

Complexity (6.1) is greater than complexity (6.2) if

$$(1 - t/m) \cdot 2^{-t} > 2^{-(p+n+1)}. \qquad (6.3)$$

If $t \ll m$ and $t < p + n + 1$ relation (6.3) holds true.

Obviously, this complexity measure only takes into account the numbers of outputs and inputs of the combinational blocks and not the functions they actually implement.

In the case of FPGA, combinational circuits are synthesized with programmable blocks called lookup tables (LUTs) which are able to implement any k-variable switching function. Typical values of current FPGAs are $k = 4$ and $k = 6$. This suggests another generic complexity measure: for that, the following property can be used:

The maximum number of external inputs of a circuit made up of $s \cdot k$-input LUTs is $s \cdot k - (s - 1)$.

It can easily be demonstrated by induction:

- If $s = 1$, the maximum number of inputs is $k = 1 \cdot k - (1 - 1)$.

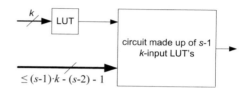

Fig. 6.2 Circuit with s k-input LUTs

- If $s > 1$, a circuit that consists of s LUTs can be decomposed into a circuit that includes $s - 1$ LUTs plus an additional LUT (Fig. 6.2); the maximum numbers of inputs is $k + [(s - 1) \cdot k - (s - 2) - 1] = s \cdot k - (s - 1)$.

Thus, the minimum number s of LUTs for implementing a function of r variables satisfies the following relation: $s \cdot k - (s - 1) \geq r$, so that

$$s \geq \lceil (r - 1)/(k - 1) \rceil. \qquad (6.4)$$

A generic measure of the delay can also be defined, based on the following property:

the maximum number of external inputs of a circuit whose critical path includes l k-input LUTs is k^l.

To check the preceding property, just observe that a tree of k-input LUTs with depth equal to l has at most k^l inputs.

Thus, if a circuit has r external inputs, the critical path includes at least l LUTs where $k^l \geq r$, so that $l \geq \log_k r$ and the circuit delay t satisfies

$$t \geq \lceil log_k r \rceil \cdot T_{LUT}, \qquad (6.5)$$

being T_{LUT} the delay of a k-input LUT.

With the maximum cost and minimum delay definitions (6.4) and (6.5), assuming that no LUT is shared by two or more functions, the maximum cost c_1 and minimum delay t_1 of the circuit of Fig. 6.1a are

$$c_1 = (m+1).\lceil (p+n)/(k-1) \rceil \text{LUTs and } t_1$$
$$= \lceil log_k(p+n) \rceil \cdot T_{LUT}. \qquad (6.6)$$

With the same definitions, the maximum cost c_2 and delay t_2 of the circuit of Fig. 6.1b are

$$c_2 = (t+1) \cdot \lceil (p+n)/(k-1) \rceil + m.\lceil (t-1)/(k-1) \rceil \text{LUTs} \qquad (6.7)$$

and

$$t_2 = (\lceil log_k(p+n) \rceil + \lceil log_k t \rceil) \cdot T_{LUT}. \qquad (6.8)$$

Example 6.1 Consider the circuit of Sect. 2.5 (*scalar_product.vhd*, available at the Authors' web site). The operations that the data path executes correspond to combinations of values of signals

start_mult, load, shift, en_X$_A$, en_X$_B$, en_Z$_A$, en_Z$_B$, en_R, sel_p1, sel_p2, sel_a1, sel_a2, sel_sq, sel_x$_A$, sel_x$_B$, sel_z$_A$, sel_z$_B$, sel_R;

in total twenty-seven bits (nine 1-bit signals and nine 2-bit signals). The program execution control is based on the values of three status signals (flags) *count, msb_k, mult_done* generated by the data path plus an external *start* command. The finite-state machine (Algorithm 2.5) has forty-three states that can be encoded with six bits. Thus, $m = 27$, $p = 3$ and $n = 6$. On the other hand, the number s of different commands can be deduced from Algorithm 2.4; the following table defines for every instruction number (from 0 to 42) the command transmitted to the data path as well as a name (mnemonic) that permits to identify the command.

There are $20 \ll 2^{27}$ different commands that can be encoded with $t = 5$ bits.

Thus, the complexities in numbers of stored bits (6.1 and 6.2) are

$$(m+1) \cdot 2^{p+n+1} = 28 \cdot 2^{10}$$
$$= 28,672 \text{ and } (t+1)$$
$$\cdot 2^{p+n+1} + m \cdot 2^t$$
$$= 7,008, \qquad (6.9)$$

and the complexities in numbers of LUTs (6.6 and 6.7), assuming that 4-input LUTs are used ($k = 4$), are

$$c_1 = (m+1).\lceil (p+n)/(k-1) \rceil = 28 \cdot \lceil 9/3 \rceil$$
$$= 84 \text{ LUT's} \qquad (6.10)$$

and

$$c_2 = (t+1) \cdot \lceil (p+n)/(k-1) \rceil + m.\lceil (t-1)/(k-1) \rceil$$
$$= 6 \cdot 3 + 27 \cdot \lceil (4/3) \rceil = 72 \text{ LUT's}. \qquad (6.11)$$

The corresponding minimum delays (6.6 and 6.7) are

$$t_1 = \lceil log_k(p+n) \rceil \cdot T_{LUT}$$
$$= \lceil log_4 9 \rceil \cdot T_{LUT} = 2 \cdot T_{LUT} \qquad (6.12)$$

and

$$t_2 = (\lceil log_k(p+n) \rceil + \lceil log_k t \rceil) \cdot T_{LUT}$$
$$= (\lceil log_4 9 \rceil + \lceil log_4 5 \rceil) \cdot T_{LUT} = 4 T_{LUT}. \qquad (6.13)$$

The second complexity measure (number of LUTs) is surely more accurate than the first one. Thus, according to (6.9–6.13), the encoding of the commands hardly reduces the cost and increases the delay. So, in this particular case, the only advantage (if any) is clarity, flexibility and ease of debugging and not cost reduction.

A new version of the circuit of Sect. 2.5 has been generated. The VHDL model *scalar_product_decoder.vhd* is available at the Authors' web site. The data path is unchanged. The control unit includes the following type definition:

```
type instruction_set is (
nop, sw_reset, multAB1, squareB, updateR,
multBA1, updateB, multAA, squareA, upda-
teA,
```

multRB, multPA, updateAB, multBA2, mul-
tAB2,

multBB, multRA, multPB, updateBA, inc);

and a signal *mnemonic* of type *instruction_set* is declared. The control unit includes a block *command decoder* that associates to every value of signal *mnemonic* the value of the control signals:

```
decoder: PROCESS(mnemonic)
BEGIN
  CASE mnemonic IS
  WHEN nop =>
      start_mult <= '0'; load <= '0'; shift <= '0'; en_xA <= '0';
      en_xB <= '0'; en_zA <= '0'; en_zB <= '0';
      en_R <= '0'; sel_p1 <= "00"; sel_p2 <= "00"; sel_a1 <= "00";
      sel_a2 <= "00"; sel_sq <= "00"; sel_xA <= "00";
      sel_xB <= "00"; sel_zA <= "00"; sel_zB <= "00"; sel_R <= '0';
  WHEN sw_reset =>
      start_mult <= '0'; load <= '1'; shift <= '0'; en_xA <= '0';
      en_xB <= '0'; en_zA <= '0'; en_zB <= '0';
      en_R <= '0'; sel_p1 <= "00"; sel_p2 <= "00"; sel_a1 <= "00";
      sel_a2 <= "00"; sel_sq <= "00"; sel_xA <= "00";
      sel_xB <= "00"; sel_zA <= "00"; sel_zB <= "00"; sel_R <= '0';
  WHEN multAB1 =>
      start_mult <= '1'; load <= '0'; shift <= '0'; en_xA <= '0'; en_xB
      <= '0'; en_zA <= '0'; en_zB <= '1';
      en_R <= '0'; sel_p1 <= "00"; sel_p2 <= "01"; sel_a1 <= "00";
      sel_a2 <= "01"; sel_sq <= "00"; sel_xA <= "00";
      sel_xB <= "00"; sel_zA <= "00"; sel_zB <= "00"; sel_R <= '0';
  ................ .
  WHEN inc =>
      start_mult <= '0'; load <= '0';
      if count < m-1 then shift <= '1'; else shift <= '0'; end if;
      en_xA <= '0'; en_xB <= '0'; en_zA <= '0'; en_zB <= '0';
      en_R <= '0'; sel_p1 <= "00"; sel_p2 <= "00"; sel_a1 <= "00";
      sel_a2 <= "00"; sel_sq <= "00"; sel_xA <= "00";
      sel_xB <= "00"; sel_zA <= "00"; sel_zB <= "00"; sel_R <= '0';
END CASE;
END PROCESS decoder;
```

and the block that generates encoded commands is defined as follows:

```
CASE current_state IS
  WHEN 0 to 1 => mnemonic <= nop; done <= '1';
  WHEN 2 => mnemonic <= sw_reset; done <= '0';
  WHEN 3 => mnemonic <= nop; done <= '0';
  WHEN 4 => mnemonic <= multAB1; done <= '0';
  .....
     WHEN 41 => mnemonic <= updateBA;
done <= '0';
  WHEN 42 => mnemonic <= inc; done <= '0';
END CASE;
```

Comment 6.1

The use of encoded commands and of mnemonics is a common practice in microprocessor programming: every machine language instruction has a name (mnemonic) that the programmer will use even if it defines a program at this low level. In this case, a set of instructions (Table 6.1) has been defined, each of them with a name (mnemonic), and to each instruction correspond a set of data path operations. The translation of the mnemonic to a vector of command signals is executed by a command decoder circuit (Fig. 6.1b). Once this decoder has been defined, the generation of the control unit can be done with mnemonics instead of command signal values, making the circuit description clearer and easier to debug and modify.

Actually, there are other commonly used programming language techniques that can be used to synthesize control units. Some of them will be seen in the next sections.

Table 6.1 Commands

Instruction number	Command	Mnemonic
0,1,3,7,10,14,17, 20,26,29,33,36, 39,42 (count = m-1)	no operation	nop
2	$x_A = 1$, $z_A = 0$, $x_B = x_P$, $z_B = 1$, count = 0	sw_reset
4	$z_B = x_A + z_A$, start($Z = x_A \cdot z_B$)	multAB1
5,6,32	$z_B = z_B^2$	squareB
8,27	R = Z	updateR
9	start ($Z = x_B \cdot z_A$)	multBA1
11,18,34,40	$x_B = Z$	updateB
12	$z_A = R + x_B$, start ($Z = x_A \cdot z_A$)	multAA
13,24,25	$z_A = z_A^2$	squareA
15,21,30,37	$x_A = Z$	updateA
16	$R = x_A^2$, start($Z = R \cdot x_B$)	multRB
19	start ($Z = x_P \cdot z_A$)	multPA
22	$(x_A, z_A, x_B, z_B) = (z_B, R, x_A + x_B, z_A)$	updateAB
23	$z_A = x_B + z_B$, start($Z = x_B \cdot z_A$)	multBA2
28	start ($Z = x_A \cdot z_B$)	multAB2
31	$z_B = R + x_A$, start ($Z = x_B \cdot z_B$)	multBB
35	$R = x_B^2$, start($Z = R \cdot x_A$)	multRA
38	start ($Z = x_P \cdot z_B$)	multPB
41	$(x_B, z_B, x_A, z_A) = (z_A, R, x_B + x_A, z_B)$	updateBA
42 (count < m −1)	count = count + 1	inc

6.2   Hierarchical Control Unit

As already pointed out above, hierarchical design is a usual strategy in many fields of system engineering: hierarchy improves clarity, security, easiness to debug and to maintain, thus reducing development times.

Nevertheless, in the case of digital circuits, the use of previously defined components sometimes prevents the designer from sharing computation resources between several components. In such a case, a conventional (flat) structure could be considered. In order to maintain some type of hierarchy (meaning clarity, security and so on), the corresponding control unit could be decomposed into a main control unit, in charge of linking together the operations, and secondary control units in charge of controlling subsets of operations.

Consider an example. The diagram of Fig. 6.3 includes two PID controllers that control two physical systems (processes). As a simple PID controller has already been developed (Sect. 5.4), a straightforward solution is two instantiate two PID controllers, each of them with its corresponding coefficient values (K_{11}, K_{21}, K_{31}) and (K_{12}, K_{22}, K_{32}).

The complete circuit includes two computation resources, in particular two 2's complement multipliers and two register files. Another option is to define a data path that permits to execute both PID loops. Assuming that both PID loops have the same sampling frequency equal to $1/T$, the algorithm to be executed is the following:

Algorithm 6.1 Double PID controller

```
loop
    e₁ = r₁ - s₁;
    acc = u₁d + K₁₁·e₁;
    acc = acc + K₂₁·e₁d;
    u₁ = acc + K₃₁·e₁dd,  u₁d = acc + K₃₁·e₁dd;
    e₁dd = e₁d;
    e₁d = e₁;
    e₂ = r₂ - s₂;
    acc = u₂d + K₁₂·e₂;
    acc = acc + K₂₂·e₂d;
    u₂ = acc + K₃₂·e₂dd,  u₂d = acc + K₃₂·e₂dd;
    e₂dd = e₂d;
    e₂d = e₂;
    wait until time_out = 1;
end loop;
```

A data path able to execute Algorithm 6.1 is shown in Fig. 6.4.

- The sixteen-word register file stores the algorithm variables e_1, e_{1d}, e_{1dd}, u_{1d}, e_2, e_{2d}, e_{2dd}, u_{2d} and acc at addresses 0, 1, 2, 3, 8, 9, 10, 11 and 15.
- The programmable computation resource computes $z = x - y$, $z = x + k_{11} \cdot y$, $z = x + k_{21} \cdot y$, $z = x + k_{31} \cdot y$, $z = x + k_{12} \cdot y$, $z = x + k_{22} \cdot y$, $z = x + k_{32} \cdot y$ and $z = x$, under the control of an *oper* signal (Table 6.2).
- There are two output registers, one for each PID loop.

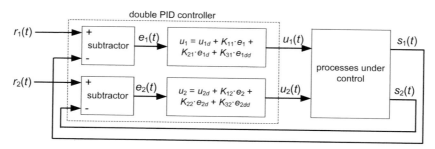

Fig. 6.3 Double PID controller

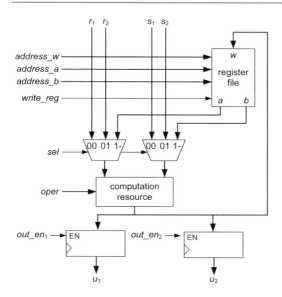

Fig. 6.4 Double PID controller: data path

Table 6.2 Computation resource operations

oper	z
000	x
010	$x + k_{11} \cdot y$
100	$x + k_{21} \cdot y$
110	$x + k_{31} \cdot y$
001	$x - y$
011	$x + k_{12} \cdot y$
101	$x + k_{22} \cdot y$
111	$x + k_{32} \cdot y$

The control unit is a thirteen-state machine whose internal states correspond to the Algorithm 6.1 instructions:

```
 0: e₁ = r₁ - s₁;
 1: acc = u₁d + K₁₁·e₁;
 2: acc = acc + K₂₁·e₁d;
 3: u₁ = acc + K₃₁·e₁dd, u₁d = acc + K₃₁·e₁dd;
 4: e₁dd = e₁d;
 5: e₁d = e₁;
 6: e₂ = r₂ - s₂;
 7: acc = u₂d + K₁₂·e₂;
 8: acc = acc + K₂₂·e₂d;
 9: u₂ = acc + K₃₂·e₂dd, u₂d = acc + K₃₂·e₂dd;
10: e₂dd = e₂d;
11: e₂d = e₂;
12: wait until time_out = 1;
```

A complete VHDL model *double_pid_controller.vhd* is available at the Author's web site.

As quoted above, in order to maintain some type of hierarchy, the corresponding control unit could be decomposed into a main control unit and two secondary control units. The control unit structure is shown in Fig. 6.5. The next-state and output functions are shown in Fig. 6.6 and Tables 6.3 and 6.4.

The first secondary control unit generates the control signals that correspond to the first PID loop execution, and the second secondary control unit generates the control signals that correspond to the second PID loop execution. In both secondary units, the control signal values corresponding to *nop* (no operation) are *write_reg* = 0, *sel* = 00, *address_W* = 0000, *address_A* = 0000, *address_B* = 0000, *oper* = 000, so that the actual command transmitted to the data path can be generated by ORing the commands generated by both control units: when a secondary unit generates commands different from *nop*, the other unit is waiting for *start* = 1 and generates the *nop* command.

A complete VHDL model *double_pid_controller2.vhd* is available at the Author's web site. Simulation results are shown in Fig. 6.7. The first PID controller is the same as in Chap. 5 (Table 5.4 and Fig. 5.32). The second controller parameters are

$K_{12} = 73$, $K_{22} = -86$, $K_{32} = 201$, so that $u_2 = u_{2d} + 73 \cdot (r_2 - s_2) - 86 \cdot e_{2d} + 201 \cdot e_{2dd}$.

The first computation steps are shown in Table 6.5.

The proposed solution is similar to the use of procedures in programming languages: a main control unit (a main program) calls two secondary control units (two procedures) that correspond to PID loops 1 and 2.

Another solution can be considered: instead of two secondary control units, a single secondary unit is defined. The main control unit calls this secondary unit and transmits parameters corresponding to either PID loop 1 or 2. The parameters transmitted to the secondary unit are *addr_e*, *addr_e_d*, *addr_e_dd*, *addr_u_d* and *pid_number*. They are the register file addresses where e_i, e_{id}, e_{idd} and u_{id} (i = 1 or 2) must be stored plus a binary identifier of the PID loop (1 or 2) to be executed. The control unit structure is shown in Fig. 6.8.

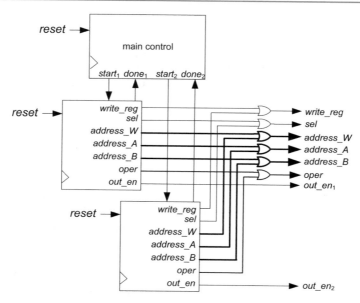

Fig. 6.5 Hierarchical control unit: structure

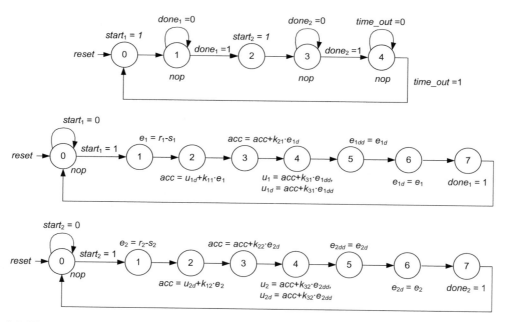

Fig. 6.6 Hierarchical control unit: next-state and output functions

Table 6.3 Signals generated by the first secondary control unit

operation	wr_reg	sel	addr_w	addr_a	addr_b	oper	en_out
$e_1 = r_1 - s_1$	1	00	0	–	–	001	0
$acc = u_{1d} + K_{11} \cdot e_1$	1	11	15	3	0	010	0
$acc = acc + K_{21} \cdot e_{1d}$	1	11	15	15	1	100	0
$u_1 = u_{1d} = acc + K_{31} \cdot e_{1dd}$	1	11	3	15	2	110	1
$e_{1dd} = e_{1d}$	1	11	2	1	–	000	0
$e_{1d} = e_1$	1	11	1	0	–	000	0
nop	0	00	0	0	0	000	0

Table 6.4 Signals generated by the second secondary control unit

operation	wr_reg	sel	addr_w	addr_a	adds_b	oper	en_out
$e_2 = r_2 - s_2$	1	01	8	–	–	001	0
$acc = u_{2d} + K_{12} \cdot e_2$	1	11	15	11	8	011	0
$acc = acc + K_{22} \cdot e_{2d}$	1	11	15	15	9	101	0
$u_2 = u_{2d} = acc + K_{32} \cdot e_{2dd}$	1	11	11	15	10	111	1
$e_{2dd} = e_{2d}$	1	11	10	9	–	000	0
$e_{2d} = e_2$	1	11	9	8	–	000	0
nop	0	00	0	0	0	000	0

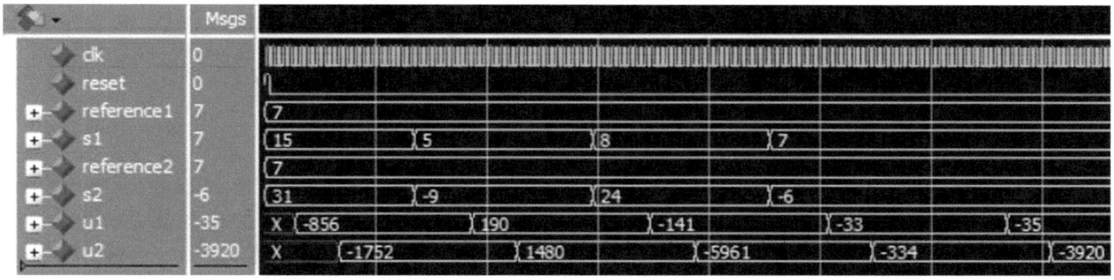

Fig. 6.7 Simulation results (courtesy of Mentor Graphics)

Table 6.5 First computation steps

r_2	s_2	u_{2d}	e_2	e_{2d}	e_{2dd}	u_2
7	31	0	−24	0	0	−1752
7	−9	−1752	16	−24	0	1480
7	24	1480	−17	16	−24	−5961
7	−6	−5961	13	−17	16	−334
7	−6	−334	13	13	−17	−3920

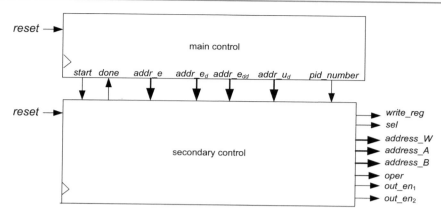

Fig. 6.8 Hierarchical control unit, version 2

The next-state functions are shown in Fig. 6.9, and the output functions are defined in Tables 6.6, 6.7 and 6.8.

A complete VHDL model *double_pid_controller3.vhd* is available at the Author's web site.

As in the previous example (Fig. 6.5), the method used in this second example of hierarchical control unit is similar to the use of procedures in programming languages. There is a difference with the first example:

- In the first example (Fig. 6.5), there are two secondary units; one for the first PID controller and another for the second controller; the first secondary unit processes data stored at addresses 0–3, and the others process data stored at addresses 8–11; the main control

unit function is to alternatively call the secondary units.

- In the second example (Fig. 6.8), there is only one secondary unit; the main control unit function is to call the secondary unit and to transmit the addresses of the data processed by either the first (addresses 0–3) or the second (addresses 8–11) PID loop; this is similar to the passing of parameters by reference in programming languages.

Comment 6.2

The definition of hierarchical control units is a technique similar to the use of procedures and functions in software generation. This type of

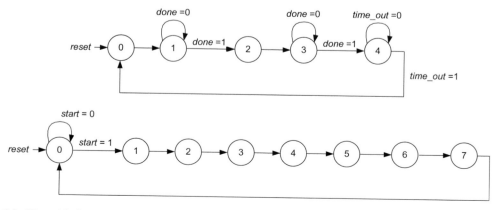

Fig. 6.9 Hierarchical control unit, version 2: next-state functions

Table 6.6 Output function of the main control unit

state	addr_e	addr_e$_d$	addr_e$_{dd}$	addr_u$_d$	pid_number	start
0	0	1	2	3	0	1
1	0	1	2	3	0	0
2	8	9	10	11	1	1
3	8	9	10	11	1	0
4	0	0	0	0	0	0

Table 6.7 Output function of the secondary control unit

state	write_reg	sel$_1$	sel$_0$	address_W	address_A	address_B
0	0	0	pid_number	0	0	0
1	1	0	pid_number	addr_e	0	0
2	1	1	pid_number	acc	addr_u$_d$	addr_e
3	1	1	pid_number	acc	acc	addr_e$_d$
4	1	1	pid_number	addr_u$_d$	acc	addr_e$_{dd}$
5	1	1	pid_number	addr_e$_{dd}$	addr_e$_d$	0
6	1	1	pid_number	addr_e$_d$	addr_e	0
7	0	0	pid_number	0	0	0

Table 6.8 Output function of the secondary control unit (continued)

state	oper$_{2-1}$	oper$_0$	out_en$_1$	out_en$_2$	done
0	00	0	0	0	0
1	00	1	0	0	0
2	01	pid_number	0	0	0
3	10	pid_number	0	0	0
4	10	pid_number	1 − pid_number	pid_number	0
5	00	0	0	0	0
6	00	0	0	0	0
7	00	0	0	0	1

approach to control unit synthesis is more a question of clarity (well-structured control unit), easiness to debug and maintain, than of cost reduction (control units are not expensive).

6.3 Variable-Latency Operations

In Sect. 2.2, operation scheduling was performed assuming that the computation times t_{JM} of all operations were constant values. Once an operation schedule has been selected, the definition of the control unit is quite obvious. Nevertheless, in some cases, the computation time is not a constant but a data-dependent value and the control unit synthesis is not so clear.

Consider an example of variable-latency operation.

Example 6.2 Design a circuit that computes y^x mod m, where x, y and m are naturals. It is the basic function of the *RSA* public key encryption algorithm (Rivest et al. 1978). If $x = x_0 + x_1 \cdot 2 + x_1 \cdot 2^2 + \cdots + x_{k-1} \cdot 2^{k-1}$ then

$$y^x = y^{x_0 + x_1 \cdot 2 + x_2 \cdot 2^2 + \cdots + x_{k-1} \cdot 2^{k-1}}$$

$$= y^{x_0} \cdot \left(y^2\right)^{x_1} \cdot \left(y^{2^2}\right)^{x_2} \cdot \ldots \cdot \left(y^{2^{k-1}}\right)^{x_{k-1}} \quad (6.14)$$

so that the following algorithm computes $z = y^x$ mod m.

Algorithm 6.2 Mod m exponentiation

```
a = 1; b = y;
for i in 0 to k-1 loop
  if xᵢ = 1 then a = a·b mod m; end if;
  b = b² mod m;
end loop;
z = a;
```

A data path that executes Algorithm 6.2 is shown in Fig. 6.10a. Its main component is a mod m multiplier that executes $a \cdot b$ mod m and $b \cdot b$ mod m. A VHDL model *mod_mm_multiplier.vhd* of this component, with $k = 192$, is available at the Authors' web site.

To complete the circuit implementation, a control unit must be defined. Its functions are the following:

- It generates the signals *load, en_a, en_b, sel, start_mult* and *shift* in function of x_i (the shift register serial output) and of the *mult_done* flag.
- It includes a k-state counter that detects the end of the loop execution (*for i in 0 to k-1 loop*).
- It executes a simple *start/done* communication protocol (Fig. 1.3) that permits the communication of the circuit with other components.

A complete VHDL model *mod_mm_expo-nentiation.vhd* of the exponentiation circuit, with $k = 192$, is available at the Authors' web site.

The computation time of the exponentiation circuit based on Algorithm 6.2 is data-dependent. Let $w(x)$ be the number of 1's of the binary representation of x. Then, Algorithm 6.2 executes k mod m products $b \cdot b = b^2$ and $w(x)$ mod m products $a \cdot b$, so that the total number of mod m products is equal to $k + w(x)$, a number included between k and $2 \cdot k$.

The exponentiation circuit has been simulated. A first simulation (Fig. 6.11) has been executed to check the working of the circuit: it computes $z = y^x$ mod m where $m = 2^{192-264} - 1$, $x = m$ and $y = 01111111 \cdots 1111 = [7f \cdots f]_{\text{hexadecimal}}$; as m is

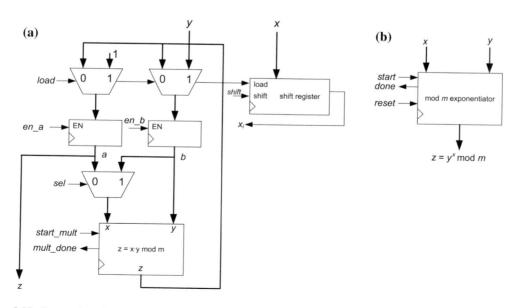

Fig. 6.10 Data path and symbol of a circuit that computes $z = y^x$ mod m

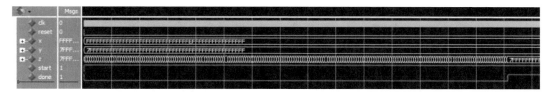

Fig. 6.11 First simulation (little Fermat's theorem) (courtesy of Mentor Graphics)

a prime number then, according to the little Fermat's theorem, $y^m \bmod m = y$.

A second simulation (Fig. 6.12) shows that this is a variable-latency operation. It computes $z = y^x \bmod m$ with the same values as before. Then, it computes $z = y^x \bmod m$ with the same values of y and m and with $x = 10000000 \cdots 0000 = [80 \; \cdots \; 0]_{\text{hexadecimal}}$. The first computation takes 75,644 cycles (from the falling edge to the rising edge of *done*), while the second takes 38,404. Actually, in the first case, $w(x) = 191$ so that the number of products is equal to $k + w(x) = 383$, while in the second case, $w(x) = 1$ so that the number of products is equal to $k + w(x) = 193$. Each $\bmod m$ product takes $k = 192$ cycles plus some additional initialization and termination cycles, so that the total numbers of cycles are greater than $383 \cdot 192 = 73{,}536$ and $193 \cdot 192 = 37{,}056$, respectively (Fig. 6.12).

Consider an algorithm that includes one or several operations whose computation time is data-dependent. The methods of Chap. 2 could still be used if upper bounds of the number of cycles necessary to complete each operation are known. In the case of the exponentiation circuit of Example 6.2, the maximum computation time is equal to $2k = 384$ cycles, plus some initialization and termination cycles, so that the computation time of this component is equal to e.g. 388 cycles. However, this could be an inefficient option as the

total algorithm execution time could be much longer than actually necessary. A better option could be a kind of dynamic operation scheduling based on the use of status signals (flags) generated by the circuits that implement the operations. As an example, the control unit of the exponentiation circuit of Example 6.2 generates an output *done* = 1 (Fig. 6.10b) when it has completed a computation and the operation result is available. So, the VHDL description of a circuit that uses the exponentiation circuit of Fig. 6.10b should probably include sentences such as (or equivalent to)

```
wait for done = '1'.
```

To summarize, in the case of variable-latency components, two options could be considered:

- A first one is to previously compute an upper bound of their computation times, if such a bound exists.
- Another option is to use a *start/done* protocol: *done* is lowered on the *start* positive edge and raised when the results are available.

The second option is more general and generates circuits whose average latency is shorter. In the particular case of pipeline circuits, an interesting implementation method has been

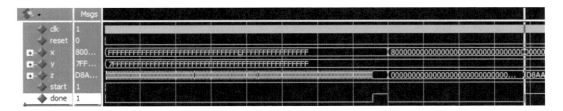

Fig. 6.12 Second simulation (computation time is data-dependent) (courtesy of Mentor Graphics)

described in Sect. 3.5: a self-timed circuit based on a handshaking protocol making use of the *start* and *done* signals of each pipeline stage (Figs. 3.32, 3.33, 3.34 and Example 3.4). Obviously handshaking protocols can be used in any circuit including variable-latency components.

Another way to control the transmission of data between variable-latency components is to use flexible inter-component connections, for example, FIFO files. In the following example, *start/done* protocol techniques and FIFO interface are used.

Example 6.3 The system of Fig. 6.13 consists of an emitter and a receiver connected by a transmission line. The emitter sends messages that are sequences of *n* encoded words. Each word is a natural and is encoded with the *RSA* publickey encryption algorithm. Given a word w_i of the plaintext message, the transmitted data is $z_i = w_i^e$ mod *m* being *e* the public key. On the receiver side, this encoded data z_i is decoded by computing z_i^d mod *m* being *d* the private key. Module *m* is a *k*-bit natural, and all constants and processed variables, namely *e*, *d*, w_i and z_i, are *k*-bit naturals smaller than *m*.

Assume that a variable-latency exponentiation algorithm (see Example 6.2) is used to compute w_i^e mod *m* (emitter) and z_i^d mod *m* (receiver). Then, it might happen that the time interval between successive data sent by the emitter to the receiver be shorter than the decoding time within the receiver. That will happen if the number of 1's of *e* is smaller than the number of 1's of *d*. For that reason, a memory is necessary on the receiver side to store the incoming data. A possible option is shown in Fig. 6.14: the transmitted data are stored in a FIFO file under the control of the *ready* signal and are read from the FIFO file under the control of an *exp_done* signal generated by the mod *m* exponentiation circuit of the receiver.

A complete VHDL model *encoded_transmission.vhd* is available at the Authors' web site.

- The emitter consists of a *signal generator* that generates random sequences of *n* *k*-bit naturals smaller than *m* and of a mod *m* exponentiation circuit. For that the functions, UNIFORM (random number generation) and TRUNC (truncation) defined within the IEEE.MATH_REAL package are used.

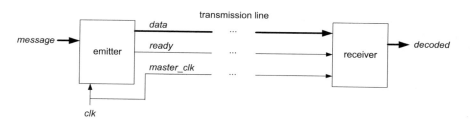

Fig. 6.13 Encoded transmission

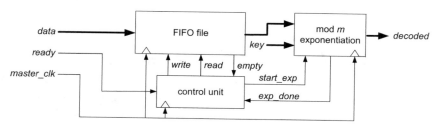

Fig. 6.14 Receiver structure

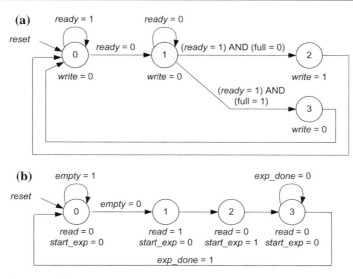

Fig. 6.15 Control unit: **a** write fifo; **b** read fifo and start exponentiation

- The receiver structure is shown in Fig. 6.14, and the control unit is described in Fig. 6.15a and b. It consists of two independent finite-state machines.

The partition of the control unit corresponds to the fact that the data input rate to the FIFO and the data output rate from the FIFO are different. The input rate is determined by the emitter, and the output rate is determined by the decoding exponentiation circuit.

The use of a FIFO with separate input and output control units is an example of flexible inter-component connection.

A simulation result is shown in Fig. 6.16. It corresponds to the transmission of $n = 32$ words; each word is an 8-bit natural ($k = 8$); $m = 91$; e (public key) is equal to 5 and d (private key) is equal to 29. The final words of the original

message and of the encoded message are shown in Fig. 6.15. The decoded message is (obviously) identical to the original message with some delay due to the difference between the computation times of w_i^5 mod 91 and z_i^{29} mod 91.

According to the conclusions of Example 6.2, the number of mod m products to execute Algorithm 6.2 is equal to k plus the number of 1's in the exponent: two 1's in the case of $e = 00000101$ and four 1's in the case of $d = 00011101$. In the first case, each exponentiation includes $8 + 2 = 10$ mod m products, and in the second case, each exponentiation includes $8 + 4 = 12$ products. Thus, the decoding operation is roughly 1.2 times slower than the encoding operation so that the decoding of a 32-bit message takes about the same times as the encoding of $32 \cdot 1.2 = 38.4$ words, roughly seven more words, a fact that is confirmed by the simulation result.

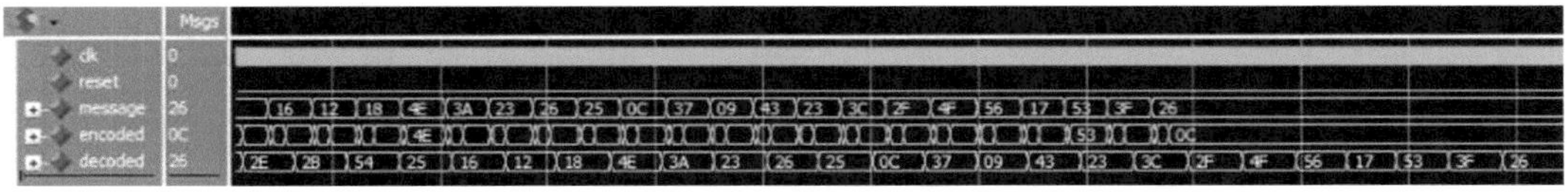

Fig. 6.16 Simulation (courtesy of Mentor Graphics)

Comment 6.3

A typical case of data-dependent computation time corresponds to algorithms that include *while* loops: some iteration is executed as long as a condition holds true. A straightforward implementation method is the use of a component executing some kind of *start/done* protocol. Now, assume that the resulting circuit is too slow. A possible option (Chap. 4) is to unroll the loop (partially or completely). For that, the *while* loop must be substituted by a *for* loop including a fixed number of steps, such as *for i in* 0 *to n* − 1 *loop*. Thus, in some cases, it might be worthwhile to substitute a variable-latency slow component that executes a *while* loop by a constant-latency component.

6.4 Sequencers and Microprograms

Control units are modeled by finite state machines. Modern EDA tools include synthesis programs that use classical techniques of finite-state machine implementation. They define an appropriate encoding of the internal states in function of optimization criteria (cost, speed, consumption). However, finite-state machines that model the control unit of a circuit that implements an algorithm have some characteristics that permit to define specific implementation structures.

Consider the generic circuit of Fig. 1.4. Assuming that a Moore model is used, the corresponding control unit structure is shown in

Fig. 6.17. It consists of two combinational circuits. One computes the next state in function of the current internal state and of input signals such as conditions (flags) generated by the data path or external commands such as a *start* signal. The other combinational circuit generates commands sent to the data path and status information transmitted to external circuits, for example, a *done* signal, in function of the current internal state (Moore model).

Assume that the internal states are represented by naturals 0, 1, 2,.... It was the case in practically all examples of this book (VHDL models available at the Authors' web site). When the finite-state machine is the control unit of a data path that executes an algorithm (a program), in many (most) cases, the next state after state number i is state number $i + 1$, independently of the input variable (*conditions* and *start* in Fig. 6.17); this situation is called *normal sequence*. In other cases, the next state after state number i is state number $j \neq i + 1$, independently of the input variable; this situation is called a *jump*. Another case is when the next state after state number i is state number $j \neq i + 1$ if some Boolean condition depending on input signal values (a data path flag, an external *start* command) is true and is state number $i + 1$ in the contrary case; this situation is called a *conditional jump*. A last case is when the next state after state number i is state number $j \neq i + 1$ if some Boolean condition is true and is state number $k \neq i + 1$ in the contrary case; this situation is called a two-*way branching*.

Fig. 6.17 Control unit (Moore model)

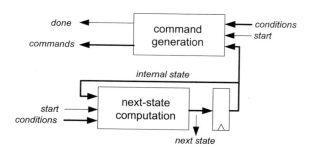

The following pieces of VHDL code correspond to those four cases.

```
--normal sequence:
CASE current_state IS
    WHEN i => current_state = current_ state +1;

--jump
CASE current_state IS
    WHEN i => current_state = j;

--conditional jump
CASE current_state IS
    WHEN i => if (condition = true) then current_state = j;
else current_state = current_state +1;

--2-way branching
CASE current_state IS
WHEN i => if (condition = true) then current_state = j;
                else current_state = k;
```

Generally, there are few two-*way branching* instructions, and if any, it can be replaced by a conditional jump followed by a jump. Obviously, there is one more instruction, and the instruction numbering must be updated.

As an example, with this type of representation, the next-state function defined by Algorithm 2.5 of Chap. 2 is equivalent to the following algorithm that includes three types of next-state computation: *normal sequence, jump* and *conditional jump.*

```
--equivalent to a 2-way branching
CASE current_state IS
WHEN i => if (condition = true) then current_state = j;
else current_state = current_state + 1;
    WHEN i+1 => current_state = k;
```

In what follows the four types of next-state definition are represented as follows:

```
--normal sequence:
i: increment;
--jump
i: goto j;
--conditional jump
i: if condition goto j;
--2-way branching
i: if condition goto j; else goto k;
```

as quoted above the preceding is equivalent to

```
--equivalent to a 2-way branching
i: if condition goto j;
i+1: goto k;
```

Algorithm 6.3 Algorithm 2.5 with three instruction types

```
0:  if start = 1 goto 0;
1:  if start = 0 goto 1;
2:  to next
3:  if msb_k = 1 goto 23;
4 to 6: to next
7:  if mult_done = 0 goto 7
8 to 9: to next
10: if mult_done = 0 goto 10
11 to 13: to next
14: if mult_done = 0 then goto 14
15 to 16: to next
17: if mult_done = 0 goto 17
18 to 19: to next
```

```
20: if mult_done = 0 goto 20
21:  to next
22: goto 42
23 to 25: to next
26: if mult_done = 0 goto 26
27 to 28: to next
29: if mult_done = 0 goto 29
30 to 32: to next
33: if mult_done = 0 goto 33
34 to 35: to next
36: if mult_done = 0 goto 36
37 to 38: to next
39: if mult_done = 0 goto 39
40 to 41: to next
42: if count = m-1 goto 0;
43: goto 3;
```

Observe that instruction number 42 of original next-state function Algorithm 2.5 has been replaced by instructions 42 and 43. Furthermore, the only instruction of the program (Algorithm 2. 4) that defines next states and operations in which the executed operation depends on an input signal value (*count*), namely

```
42: if count < m-1 then count = count
+1, go to 3;
    else go to 0;
```

has been split into two equivalent instructions

```
42: if count = m-1 then go to 0;
43: count = count + 1, go to 3;
```

so that the control unit is a Moore state machine.

Consider a control unit whose behavior is defined by an algorithm such as Algorithm 6.3 with three types of instruction types. Each instruction computes the next internal-state number in function of four pieces of information:

- current internal-state number,
- type of instruction (normal sequence, jump, conditional jump),
- value (true or false) of a condition in case of conditional jump,
- next-state number in case of jump or conditional jump.

This suggests the structure of Fig. 6.18 that consists of the following blocks:

- a combinational circuit called "control program" that associates to every internal-state number the *type* of the instruction (normal sequence, jump, conditional jump), an identifier *cond_id* of the condition to be evaluated in the case of conditional jump, and a next-state number *jump_num* in case of jump or conditional jump,
- a combinational circuit that evaluates whether a particular condition *cond* on input signal values, identified by a condition identifier, is true or false,

Fig. 6.18 Next-state computation

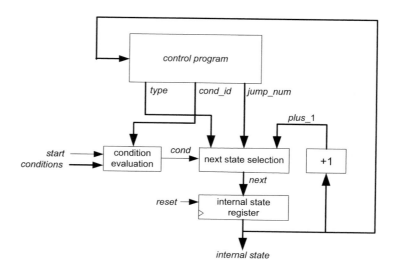

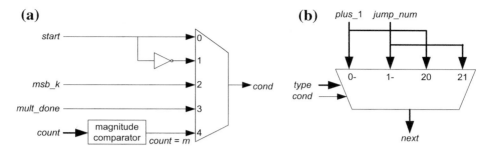

Fig. 6.19 **a** Condition evaluation. **b** Next-state number selection

- a next-state selection circuit that selects the next-state number among *plus_1* and *jump_num*,
- an increment circuit (half adder) that computes *plus_1* = current state + 1,
- an internal-state register.

Consider again Algorithm 6.3. It includes five different conditions that will be identified by numbers 0–4:

$$0 : start = 1; 1 : start = 0; 2 : msb\_k = 1; 3$$
$$: mult\_done = 0; 4 : count = m - 1;$$

The corresponding condition evaluation circuit implementation is shown in Fig. 6.19a.

Assume that the three instruction types are encoded as follows:

$$0 : \text{normal sequence}, 1 : \text{jump}; 2$$
$$: \text{conditional jump}.$$

Then, the next-state selection block works as follows (Fig. 6.19b):

- If *type* = 0, then *next* = *plus_1*.
- If *type* = 1, then *next* = *jump_num*.
- If *type* = 2 and *cond* = false, then *next* = *plus_1*.
- If *type* = 2 and *cond* = true, then *next* = *jump_num*.

The working of the "control program" circuit is defined by the following Table 6.9.

A common option is to use a read-only memory to implement the control program block defined by Table 6.9. Then, the current internal-state number is the ROM address. The ROM contents constitute a so-called *micro-program* made up of three *microinstruction* types (normal sequence, jump, conditional jump). Each microinstruction contains three *fields*: microinstruction type, condition identifier and jump address. A fourth field *command* is added: it associates to every internal-state number (microinstruction address) the corresponding values of the commands sent to the data path and status information transmitted to external circuits (Moore machine). The complete circuit is shown in Fig. 6.20. The *sequencer* block includes the condition evaluation block (Fig. 6.19a), the next-state selection block (Fig. 6.19b), the increment circuit and the internal-state register.

A complete VHDL model *scalar_product_micro.vhd* is available at the Authors' web site. To check the working of the circuit, the same values as in Chap. 2 are used:

$$x_P = 2\text{fe}13\text{c}0537\text{bbc}11\text{acaa}07\text{d}793\text{de}4\text{e}6\text{d}5\text{e}5\text{c}94\text{eee}8,$$
$$k = 4000000000000000000020108a2e0cc0d99f8a5ef,$$

and the result is

$$x_A = 1\text{d}538\text{b}8105663\text{e}13\text{c}972\text{bf}682\text{b}49975\text{f}7\text{a}5\text{fd}6345,$$
$$z_A = 4\text{ae}93681\text{fa}9\text{e}59\text{e}7\text{a}7\text{aa}2\text{b}2592\text{ba}6\text{e}92\text{dcb}7\text{d}4674,$$
$$x_B = 2758\text{e}50\text{c}38\text{d}039\text{b}358\text{daf}65\text{e}05\text{bdd}89\text{f}8\text{fb}1\text{e}4\text{a}1\text{a},$$
$$z_B = 0000000000000000000000000000000000000000.$$

Table 6.9 Control program

state number	type	cond_id	jump_num
0	2	0	0
1	2	1	1
2	0	–	–
3	2	2	23
4–6	0	–	–
7	2	3	7
8–9	0	–	–
10	2	3	10
11–13	0	–	–
14	2	3	14
15–16	0	–	–
17	2	3	17
18–19	0	–	–
20	2	3	20
21	0	–	–
22	1	–	42
23–25	0	–	–
26	2	3	26
27–28	0	–	–
29	2	3	29
30–32	0	–	–
33	2	3	33
34–35	0	–	–
36	2	3	36
37–38	0	–	–
39	2	3	39
40–41	0	–	–
42	2	4	0
43	1	–	3

Consider again Algorithm 2.4 that has been used to implement the scalar product circuit: most instructions are either a jump, conditional jump or branching, without any operation execution, or an operation execution with normal sequence. The only exceptions are

```
22: (xA, zA, xB, zB) = (zB, R, xA+xB, zA),
go to 42;
```

and

```
42: if count < m-1 then count = count +1,
go to 3;
    else go to 0;
```

Instruction 22 can be replaced by two instructions: first execute the operation

$$(x_A, z_A, x_B, z_B) = (z_B, R, x_A+x_B, z_A)$$

and then jump to the next instruction (43 after renumbering). Old instruction 42 (now instruction 43) can be replaced by three instructions: first detect the end of computation:

```
if count = m-1 then go to 0;
```

then execute the operation:

```
count = count +1;
```

finally jump to the next instruction:

```
go to 3;
```

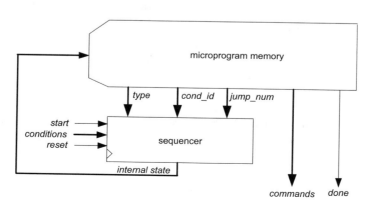

Fig. 6.20 Microprogrammed control unit

With those modifications, Algorithm 6.4, equivalent to Algorithm 2.4, is obtained. Equivalence means that the results of the algorithm executions are the same. Obviously, Algorithm 6.4 execution needs more cycles than Algorithm 2.4, so that the corresponding circuit is (a bit) slower.

Algorithm 6.4 Algorithm 2.4 modified

```
0:  wait until start = 0;
1:  wait until start = 1;
2:  xA = 1, zA = 0, xB = xP, zB = 1, count = 0;
3:    if km-i = 1 then go to 23;
4:      zB = xA+zA, start(Z = xA·zB);
5:      zB = zB²;
6:      zB = zB²;
7:      wait until mult_done = 1;
8:      R = Z;
9:      start (Z = xB·zA);
10:     wait until mult_done = 1;
11:     xB = Z;
12:     zA = R + xB, start (Z = xA·zA);
13:     zA = zA²;
14:     wait until mult_done = 1;
15:     xA = Z;
16:     R = xA², start(Z = R·xB);
17:     wait until mult_done = 1;
18:     xB = Z;
19:     start (Z = xP·zA);
20:     wait until mult_done = 1;
21:     xA = Z;
22:     (xA, zA, xB, zB) = (zB, R, xA+xB, zA);
23:     go to 43;
24:     zA = xB+zB, start(Z = xB·zA);
25:     zA = zA²;
26:     zA = zA²;
27:     wait until mult_done = 1;
28:     R = Z;
29:     start (Z = xA·zB);
30:     wait until mult_done = 1;
31:     xA = Z;
32:     zB = R + xA, start (Z = xB·zB);
33:     zB = zB²;
34:     wait until mult_done = 1;
35:     xB = Z;
36:     R = xB², start(Z = R·xA);
```

```
37:     wait until mult_done = 1;
38:     xA = Z;
39:     start (Z = xP·zB);
40:     wait until mult_done = 1;
41:     xB = Z;
42:     (xB, zB, xA, zA) = (zA, R, xB+xA, zB);
43:     if count = m-1 then go to 0;
44:     count = count +1;
45:     go to 0;
```

The interesting point is that each instruction includes either a jump (conditional or not) or an operation. This fact suggests the possibility of merging the *jump_num* and *commands* fields within the microprogram memory of Fig. 6.20. Taking into account that generally *jump_num* has fewer bits than *commands*, an interesting option is to insert a command decoder (Fig. 6.1b).

Using the same mnemonics as in Table 6.1, Algorithm 6.4 can be rewritten as follows.

Algorithm 6.5 Microprogram

```
0:  if start = 1 goto 0, done = 1;
1:  if start = 0 goto 1, done = 1;
2:  sw_reset, done = 0;
3:  if msb_k = 1 goto 24, done = 0;
4:  multAB1, done = 0;
5:  squareB, done = 0;
6:  squareB, done = 0;
7:  if mult_done = 0 goto 7, done = 0;
8:  updateR, done = 0;
9:  multBA1, done = 0;
10:   if mult_done = 0 goto 10, done = 0;
11:   updateB, done = 0;
12:   multAA, done = 0;
13:   squareA, done = 0;
14:  if mult_done = 0 then goto 14, done = 0;
15:   updateA, done = 0;
16:   multRB, done = 0;
17:   if mult_done = 0 goto 17, done = 0;
18:   updateB, done = 0;
19:   multPA, done = 0;
20:   if mult_done = 0 goto 20, done = 0;
21:   updateA, done = 0;
22:   updateAB, done = 0;
```

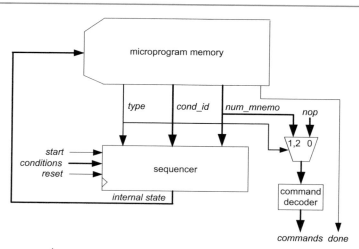

Fig. 6.21 Microprogrammed control unit: version 2

```
23:   goto 43, done = 0;
24:   multBA2, done = 0;
25:   squareA, done = 0;
26:   squareA, done = 0;
27:   if mult_done = 0 goto 27, done = 0;
28:   updateR, done = 0;
29:   multAB2, done = 0;
30:   if mult_done = 0 goto 30, done = 0;
31:   updateA, done = 0;
32:   multBB, done = 0;
33:   squareB, done = 0;
34:   if mult_done = 0 goto 34, done = 0;
35:   updateB, done = 0;
36:   multRA, done = 0;
37:   if mult_done = 0 goto 37, done = 0;
38:   updateA, done = 0;
39:   multPB, done = 0;
40:   if mult_done = 0 goto 40, done = 0;
41:   updateB, done = 0;
42:   updateBA, done = 0;
43:   if count = m-1 goto 0, done = 0;
44:   inc, done = 0;
45:   goto 3, done = 0;
```

A circuit that implements this algorithm is shown in Fig. 6.21. It consists of the following components:

- *sequencer* that includes the condition evaluation block (Fig. 6.19a), the next-state selection block (Fig. 6.19b), the increment circuit and an internal-state register,
- *command decoder* that associates to every value of signal *mnemonic*, represented by a 5-bit vector, the value of the control signals; the relation between mnemonics and control signals is defined in Table 6.1,
- *microprogram memory*: to each internal-state number (memory address) are associated four pieces of information—an instruction type *u_type*, a condition identifier *cond_id*, a field *num_mnemo* equal to the jump address (6-bit number) in case of *jump* or *conditional jump* instructions and equal to an encoded command (5-bit number) in case of *normal sequence* instruction,
- a multiplexer that transmits the command *nop* to the decoder in case of *normal sequence* instruction.

The microprogram memory contents are directly deduced from Algorithm 6.5. The corresponding VHDL code is a CASE construct that defines the microprogram memory contents:

```
CASE integer_address IS
    WHEN 0 => u_type <= "10"; cond_id <= "000"; num_mnemo <= "000000";   done <= '1';
    WHEN 1 => u_type <= "10"; cond_id <= "001"; num_mnemo <= "000001";   done <= '1';
    WHEN 2 => u_type <= "00"; cond_id <= "000"; num_mnemo <= '0'&sw_reset; done <= '0';
    WHEN 3 => u_type <= "10"; cond_id <= "010"; num_mnemo <= "011000";   done <= '0';
    WHEN 4 => u_type <= "00"; cond_id <= "000"; num_mnemo <= '0'&multAB1; done <= '0';
    WHEN 5 => u_type <= "00"; cond_id <= "000"; num_mnemo <= '0'&squareB; done <= '0';
    ........
    WHEN 42 => u_type <= "00"; cond_id <= "000"; num_mnemo <= '0'&updateBA; done <= '0';
    WHEN 43 => u_type <= "10"; cond_id <= "100"; num_mnemo <= "000000";   done <= '0';
    WHEN 44 => u_type <= "00"; cond_id <= "000"; num_mnemo <= '0'&inc;    done <= '0';
    WHEN 45 => u_type <= "01"; cond_id <= "000"; num_mnemo <= "000011";   done <= '0';
END CASE;
```

In fact, the *microprogram memory* component is a combinational circuit with 6 inputs (internal state) and 12 outputs (*u_type*: 2 bits, *cond_id*: 3 bits, *num_mnemo*: 6 bits, *done*: 1 bit) that can be synthesized with components of any available cell library (gates, LUTs, PLDs and so on), not necessarily with a ROM block.

A complete VHDL model *scalar_product_decoder_micro.vhd* is available at the Authors' web site.

Comment 6.4

- As mentioned at the beginning of this section, control units are modeled by finite-state machines and modern EDA tools include synthesis programs able to generate optimized state machine implementations. Furthermore, within complex circuits, control units do not constitute the most expensive part, but perhaps the most difficult to debug, to modify and to document. So, the main advantage (if any) of a structure like that of Fig. 6.21 is clarity. All the information necessary to implement the circuit is included within an algorithm similar to Algorithm 6.5. The synthesis of the sequencer and of the command decoder is straightforward: the main task is to make a list of all jump conditions and of all data path operations. The synthesis of the *microprogram memory* block consists in translating an algorithm such as Algorithm 6.5 into a table that defines a combinational circuit. Then, any type of combinational circuit implementation can be considered. Thus, the main design effort is the debugging of the initial algorithm (Algorithm 6.5 in the preceding example).

- Additional instruction types and more sophisticated sequencers could be (and have been) considered. For example, subroutine calls (at control unit level) could be useful if identical sequences of commands must be executed from different places of the control program.

6.5 Exercises

1. Design several 8-channel PID controllers, all with the same sample period T. Generate VHDL models of all controllers.
2. Define microprogrammed implementations of Algorithm 6.1 (double PID controller).

Bibliography

De Micheli G (1994), Synthesis and Optimization of Digital Circuits. McGraw-Hill, New York.

Rivest R, Shamir A, Adleman L (1978), A Method for Obtaining Digital Signatures and Public-Key Cryptosystems, Communications of the ACM 21 (2).

The way digital circuit components communicate between them has already been dealt with in the preceding chapters. Many examples have been proposed in which a *start/done* protocol is used to permit the communication with other components. A more complete handshaking protocol has also been defined and implemented in Sect. 3.5. Sequential implementations of connections have been considered in Sect. 5.1 (data path connectivity); they are based on the use of multiplexers or of equivalent internal buses.

This chapter deals with the communication between digital circuits that are components of a complete system. These components could be general-purpose processors, application-specific components, memories, electronic interfaces of electromechanical devices (e.g., disk drivers), digital-to-analog and analog-to-digital converters and many others.

A commonly used technique for interconnecting such components is the use of external buses. So, the main section of this chapter (Sect. 7.2) is dedicated to the definition and description of several types of buses. A first section is dedicated to general concepts.

7.1 General Concepts

The communication between components of a system can be implemented in different ways. This section briefly describes and classifies several communication implementation techniques.

A first category of communication implementation technique is based on the use of handshaking protocols. Consider two components A and B. A basic communication implementation is shown in Fig. 7.1. To send *data* from A to B

- A puts *data* on the output port *data_out* and raises the output port *out_valid*.
- When B has read *data,* it raises the output port *in_ack.*
- A lowers the output port *out_valid.*
- B lowers the output port *in_ack.*

More sophisticated handshake protocols can be considered. An example has been given in Sect. 3.5 (self-timed circuits). Several (more than one) data can be transmitted with this type of protocol. Then *valid* and *ack* signals are associated with every *data_out* port.

A frequent case is the communication between a processor and a memory. In Fig. 7.2, component A is a processor and component B a memory. An *address_valid* signal indicates that a memory operation (*read* or *write*) must be executed, and a *write* signal defines the particular operation (0: *read*, 1: *write*). Practical implementation of this type of communication protocol, namely synchronous buses protocol with an address/data strobe signal *ADS* used as an *address_valid* signal, is described in Sect. 7.2.

In the case of address-less memories (FIFO, FILO) the *address_valid* control signal generated by component A in Fig. 7.2 is replaced by two

© Springer Nature Switzerland AG 2019
J.-P. Deschamps et al., *Complex Digital Circuits*,
https://doi.org/10.1007/978-3-030-12653-7_7

Fig. 7.1 Simple handshaking

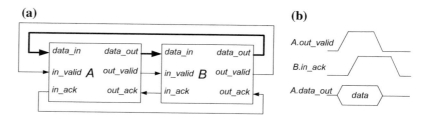

Fig. 7.2 Communication with a memory

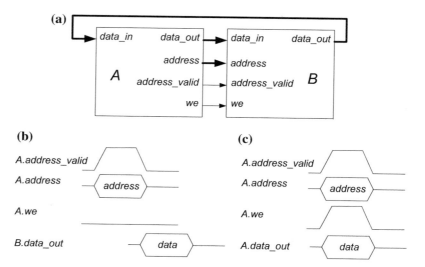

current memory status flags *not_empty* and *not_full* generated by the memory. In Fig. 7.3, component *A* is a processor and component *B* an address-less memory, for example, a first-in first-out file. A *read* operation is possible only if *not_empty* is true, and a *write* operation is possible only if *not_full* is true.

First-in first-out (FIFO) memories can be used to implement flexible inter-component connections. An example of data transmission from a component *A* to a component *C* through an address-less memory *C* is shown in Fig. 7.4. A practical case has been defined and implemented in Sect. 6 (Example 6.2).

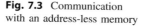

Fig. 7.3 Communication with an address-less memory

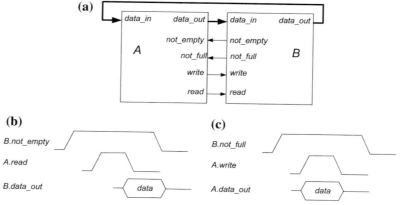

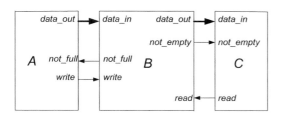

Fig. 7.4 Communication through a FIFO

In the case of systems consisting of several (more than two) components, the most commonly used technique is the use of external buses. For that reason, the second section of this chapter is dedicated to the definition and description of several types of buses.

7.2   Buses

The structure of a generic system consisting of several components is shown in Fig. 7.5. In order to interchange data, the components use a common shared connection resource called *bus* based on the use of three-state buffers (Sect. 2.4.3 of Deschamps et al. 2017). Apart from the type of data that can be transferred (basically the number of bits per word), every bus has particular and essential characteristics. For example, "how does the bus control the transfer of data between two

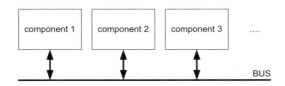

Fig. 7.5   Generic system

components?", or "what components have access to the bus at any moment?" Several particular bus types are described in the next sections.

7.2.1   Synchronous Bus

A first and very simple example of digital circuit configuration is shown in Fig. 7.6. One of the components is the *bus master* while the other components are *bus slaves*. All data transfers are between the master and one of the slaves: the master component can receive a data stored in one of the slaves (a *read* operation) and can transfer a data to one of the slaves (a *write* operation). The bus consists of the following signals:

- a synchronization signal *clk*,
- an address/data strobe *ADS*,
- a *write* signal,
- a *data* bus that transfers data,
- an *address* bus that sends addresses to the slave components.

Assume that there are four slave components and that a 16-bit address bus $a_{15} \, a_{14} \, \ldots \, a_0$ is used. Then, the two most significant bits $a_{15} \, a_{14}$ of the address can be used to select one of the slaves and the remaining bits $a_{13} \, a_{12} \, \ldots \, a_0$ address a data within the selected slave. Chronograms of *read* and *write* operations are shown in Fig. 7.7. In the first case (read operation, Fig. 7.7a), the master reads the data stored at address $a_{13} \, a_{12} \, \ldots \, a_0$ of slave number 2. The complete read operation consists of two clock cycles:

Fig. 7.6   Synchronous bus

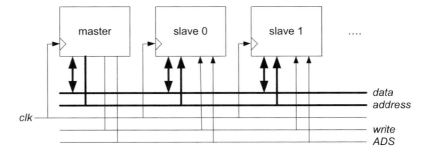

Fig. 7.7   Bus operations: **a** read and **b** write

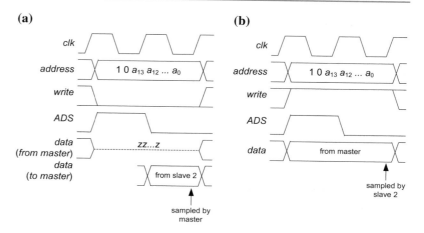

- The master generates a 16-bit address that remains stable during two cycles.
- The master generates a write bit equal to 0 (inactive) during two cycles.
- The master generates an *ADS* bit equal to 1 during the first cycle.
- The master data output is in high impedance state (ZZ ... Z) during two cycles.
- The selected slave is assumed to output a data during the second cycle; this data is sampled by the master at the end of the second cycle.
- The data output of all other slaves is in high impedance state.

The write operation (Fig. 7.7b) also consists of two clock cycles:

- The master generates a 16-bit address that remains stable during two cycles.
- The master generates a write bit equal to 1 (active) during two cycles.
- The master generates an *ADS* bit equal to 1 during the first cycle.
- The master outputs the data to be transmitted during two cycles; this data is assumed to be sampled by the slave at the end of the second cycle.
- The data output of all slaves is in high impedance state.

This could be a previously defined bus, with its particular characteristics such as the data and address sizes and the communication protocol.

When designing a new specific slave component to be connected to this bus, those characteristics must be taken into account. For instance, the specific component must have a bidirectional port *data*, an input port *address*, two 1-bit input ports *write* and *ADS*; it must include an address decoder that detects when it is addressed by the bus master. On the other hand, if an already existing component is used as bus slave, an interface circuit must be inserted whose function is to translate the bus commands to the component commands. Consider an example.

Example 7.1 Consider the system of Fig. 7.6 where the bus master is a microprocessor with an 8-bit bidirectional port *data*, a 16-bit output port *address* and two 1-bit control outputs *write* and *ADS*. One of the slaves is a synchronous random access memory (Fig. 7.8) with a 14-bit input port *address*, an 8-bit input port *data_in*, an 8-bit output port *data_out* and two control inputs *WE* (write enable) and *OE* (output enable). It stores 2^{14} 8-bit words. A new word $d_7 d_6 ... d_0$ is stored at address $a_{13} a_{12} ... a_0$ on the rising edge of *clk* if $address = a_{13} a_{12} ... a_0, data\_in = d_7 d_6 ... d_0$ and $WE = 1$ (Fig. 7.8c). If $OE = 1$ and $address = a_{13} a_{12} ... a_0,$ then *data_out* is equal to the word stored at address $a_{13} a_{12} ... a_0$. If $OE = 0$, then $data\_out = ZZZZZZZZ$ (Fig. 7.8b).

To connect this component to the bus of Fig. 7.6 as slave number 2, an interface circuit must be added (Fig. 7.9): a three-state finite-state

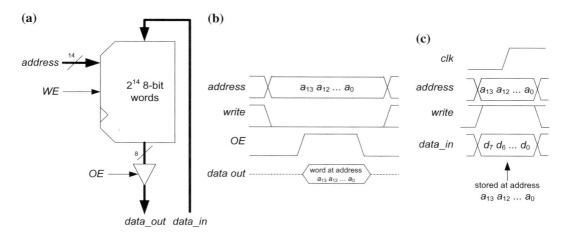

Fig. 7.8 Synchronous RAM

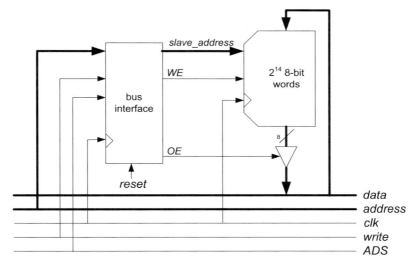

Fig. 7.9 Connection to the bus

machine generates control signals *WE* and *OE* (Fig. 7.10) and *slave_address* is equal to *address* (13 ... 0).

A VHDL model *bus_interface.vhd* is available at the Authors' web site.

In Fig. 7.7, it is assumed that when the master reads a data stored in slave number $a_{15}a_{14}$, the requested data is available on the data bus at the end of the second cycle (Fig. 7.7a). If the addressed slave is relatively slow, it could happen that it needs more than two clock cycles to put the requested data on the data bus. Similarly

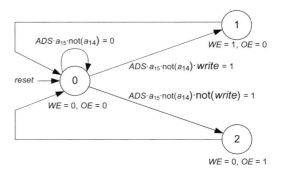

Fig. 7.10 Generation of signals *WE* and *OE*

Fig. 7.11 Synchronous bus with waiting cycles

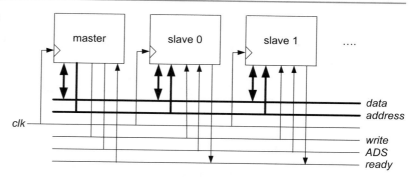

when the master sends a data to slave number $a_{15}a_{14}$ (Fig. 7.7b), it could be that more than two clock cycles are necessary to store the transmitted data within the selected slave. Slowing down the clock frequency is generally not a good solution as it will slow down all data transfers, even with "not so slow" slave components. A better option is the insertion of waiting cycles. For that, a *ready* signal is added to the bus (Fig. 7.11): *ready* is raised by the addressed slave component as soon as it has transmitted the requested data to the data bus in the case of a read operation (Fig. 7.12), or as soon as it has stored the transmitted data in the case of a write operation (Fig. 7.13). As several slaves can be connected to the bus, the state of the bus signal *ready* is the OR function of all corresponding slave signal (wired OR).

Example 7.2 As a second example, consider the system of Fig. 7.11 similar to Fig. 7.6 with an additional status signal *ready* generated by the slave components and read by the bus master. The bus master is a microprocessor with a 32-bit bidirectional port *data*, a 32-bit output port *address*, two 1-bit control outputs *write* and *ADS* and a binary input *ready*. One of the slaves is a dynamic random access memory (Fig. 7.14) with a 30-bit input port *address*, a 32-bit input port *data_in*, a 32-bit output port *data_out*, two control inputs *ME* (memory enable) and *WE* (write enable) and a control output *done*. It stores 2^{30} 32-bit words.

The write and read operations are executed as shown in Fig. 7.15. Once a command has been externally generated by raising the *ME* input and

Fig. 7.12 Read operation

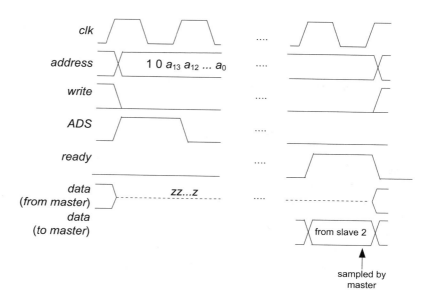

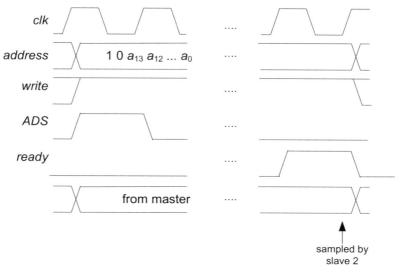

Fig. 7.13 Write operation

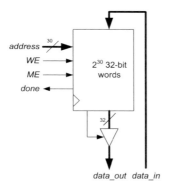

Fig. 7.14 Dynamic random access memory

by defining an address, an operation ($WE = 0$ or 1) as well as an input data in the case of a *write* operation, the internal circuits of the memory are in charge of the command execution. The *done* output is raised when either the previously addressed data is available on the output port *data_out* (*read* operation) or when the data previously inputted to the port *data_in* has been stored at the previously defined address.

To connect this component to the bus of Fig. 7.11 as slave number 2, an interface circuit must be added (Fig. 7.16): a five-state finite-state

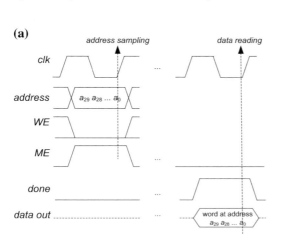

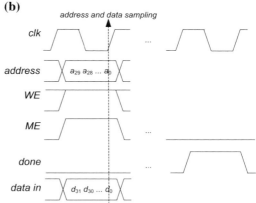

Fig. 7.15 **a** Read operation and **b** Write operation

Fig. 7.16 Connection to the bus

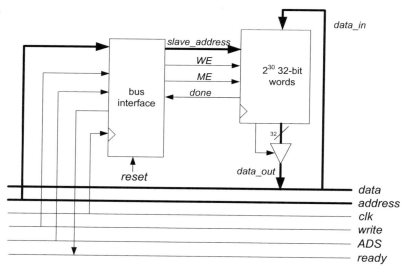

machine generates control signals *WE*, *ME* and *ready* (Fig. 7.17), and *slave_address* is equal to *address* (29 ... 0).

A VHDL model *bus_interface_2.vhd* is available at the Authors' web site.

7.2.2 Asynchronous Bus

In the preceding examples (Figs. 7.6 and 7.11), a common clock signal is used to synchronized the bus operations. Asynchronous buses without a common synchronization signal can also be defined. It is an interesting option when some of the slave components have very long response times.

An example is shown in Fig. 7.18. The bus consists of the following signals:

- An *Address/Data* bus able to transfer addresses and data from the master to the slaves and to transfer data from any slave to the master,
- *ReadReq* and *WriteReq* control bits, generated by the master, that initialize a *read* or *write* operation within the addressed slave,
- An *Ack* status bit generated by the master or by a slave to acknowledge the reception of a previous command,
- A *Ready* status bit generated by a slave when it completes the requested operation.

As before, the most significant address bits can be used to select a slave and the remaining address bits to select particular data within the selected slave. Chronograms of *read* and *write* operations are shown in Fig. 7.19a. A *read* operation consists of the following steps:

Fig. 7.17 Generation of signals *ME*, *WE* and *ready*

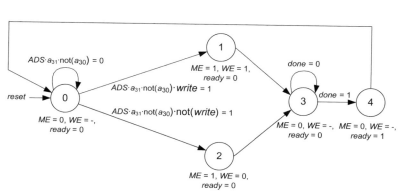

Fig. 7.18 Asynchronous bus

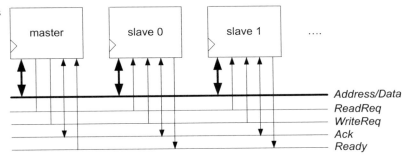

Fig. 7.19 Bus operations:
a read and **b** write

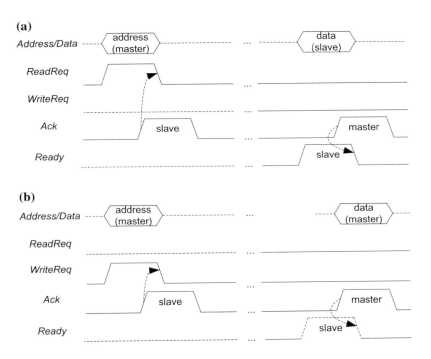

- The master puts an address on the *Address/Data* bus and raises *ReadReq*.
- The selected slave acknowledges the command by raising *Ack*.
- The master releases the *Address/Data* bus and lowers *ReadReq*.
- When ready to transmit the requested data, the previously selected slave puts the data on the *Address/Data* bus and raises *Ready*.
- The master acknowledges the reception of the requested data by raising *Ack*.
- The slave releases the *Address/Data* bus.

The write operation (Fig. 7.19b) is performed as follows:

- The master puts an address on the *Address/Data* bus and raises *WriteReq*.
- The selected slave acknowledges the command by raising *Ack*.
- The master releases the *Address/Data* bus and lowers *WriteReq*.
- When ready to store data, the previously selected slave raises *Ready*.
- The master puts the data to be stored on the *Address/Data* bus and raises *Ack*.

- The master releases the *Address/Data* bus and lowers *Ack*.

In the case of the bus of Figs. 7.11, 7.12 and 7.13, the slaves must be prepared to detect 1-cycle *ADS* strobes (all of them) and the master must be able to detect 1-cycle ready signals. Thus, a common synchronization signal is essential. This type of bus is convenient in the case of fast and close together components.

In the case of the bus of Figs. 7.18 and 7.19, simple handshaking protocols are used. For example when the master generates a *read* command (*ReadReq* = 1, Fig. 7.19a), it waits until the addressed slave acknowledges this command (*Ack* = 1). Similarly, when the addressed slave generates a *ready* pulse to indicate that a data is available on the *Address/Data* bus, it waits until the master acknowledges this

command (*Ack* = 1). Thus, the correct working of the bus does not depend on the component response delays.

Example 7.3 Consider a system that monitors the value of a set of parameters within some area: temperature, pressure, contaminant concentrations and so on. The general system structure is shown in Fig. 7.20. It consists of a central computer and of several peripherals distributed within the area to be controlled. All peripherals include sensors that measure the parameter values. The central computer executes programs that take decisions in function of those values so that it must be able to communicate with the peripherals in order to get this information. The distance between the central computer and the peripherals might be relatively long. On the other hand, this type of system does not need high-performance (speed) characteristics. So, the use of an asynchronous bus to interconnect the central computer with the peripherals could be considered (Fig. 7.18): the central computer of Fig. 7.20 is the bus master, and the peripherals are slave components.

Every peripheral has a 1-bit control input *start*, a 1-bit status output *done*, an 8-bit input port *data_in* and an 8-bit output port *data_out* (Fig. 7.21a). It works as follows (Fig. 7.21b):

- On a positive edge of *start,* the value of *data_in* is read; it identifies the particular parameter that must be detected and measured; the *done* flag is lowered.

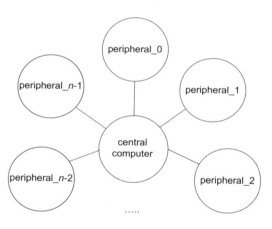

Fig. 7.20 Parameter monitoring system

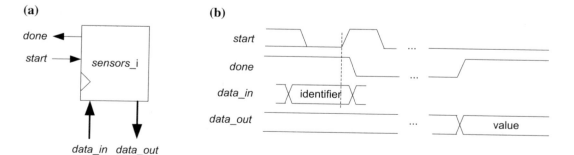

Fig. 7.21 Peripheral

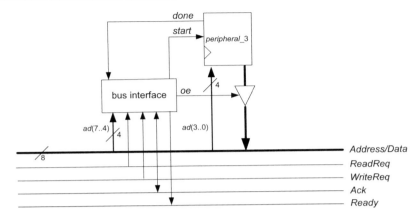

Fig. 7.22 Connection to the bus

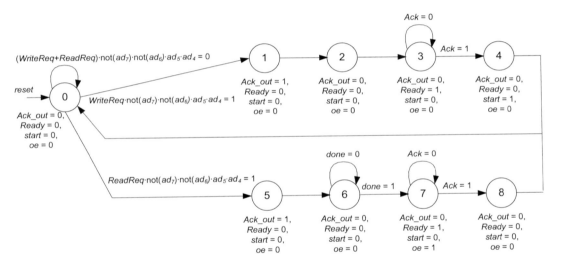

Fig. 7.23 Bus interface

- When the selected parameter has been measured, the *done* flag is raised, and the parameter value is outputted to the *data_out* port.

Assume that there are sixteen peripherals (*n* = 16 in Fig. 7.20), sixteen parameter identifiers, and that *Address/Data* (Fig. 7.18) is an 8-bit bus. To connect *peripheral_3* to the bus as slave number 3, an interface circuit is added (Fig. 7.22). When the central computer needs the value of parameter number 11 of peripheral number 3, it first executes a *write* operation (Fig. 7.19b) with

address = 0011---- and *data* = ----1011, and then, it executes a *read* operation (Fig. 7.19a) at the same address and gets the current value of the selected parameter.

The bus interface of Fig. 7.22 is a nine-state finite-state machine (Fig. 7.23) that generates signals *Ack_out, Ready_out, start* and *oe* in function of *ReadReaq, WriteReq*, bits 4–7 of *Address/Data* (peripheral identifier) and *Ack* equal to the OR function of the *Ack_out* outputs of all peripherals (wired OR).

A VHDL model *bus_interface_3.vhd* is available at the Authors' web site.

7.2.3   Multi-master Bus Systems

In the previous bus examples (Figs. 7.6, 7.11 and 7.19), there is only one bus master. All other components are slaves. Multi-master buses can also be considered. An example is shown in Fig. 7.24. All components have the capacity to control the bus operations under the supervision of an additional component called *bus arbiter*.

The way that the arbiter grants the control of the bus to a particular master is shown in Fig. 7.25:

- Master i raises the *req* output ($reqi = 1$); as the bus is currently idle, the arbiter grants the bus control to master i; master i will keep $reqi = 1$ as long as it executes bus operations.
- Masters j and k raise the corresponding *req* outputs ($reqj = reqk = 1$); as soon as master i completes its operations and lowers *reqi*, the arbiter grants the bus control to either master j or master k according to some priority policy; assume it grants the bus control to master j.
- When master j completes its operations, it lowers *reqj*; then the arbiter grants the control to master k whose request is still pending.
- When master k completes its operations, it lowers *reqk*.

Example 7.4 The system to be developed executes an algorithm of image filtering in the spatial domain. It is made up of a main processor (*cpu*)

that stores an m-row by n-column image. Each pixel is a p-bit signed number. The filtering algorithm is executed by a specific coprocessor with its own data memory. The sequence of operations executed by the main processor is the following:

- Transmission to the coprocessor data memory, at addresses a to $a + m \cdot n - 1$, of an unfiltered image.
- Transmission to the coprocessor of the following command: "execute the filtering algorithm on the image stored at addresses a to $a + m \cdot n - 1$ and store the resulting filtered image at addresses b to $b + m \cdot n - 1$" (Fig. 7.26).
- Wait for completion of the command execution; the filtered image is now stored within the coprocessor data memory.
- Reception of the filtered image from the coprocessor data memory.

Thus, the communication resource must be able to transmit data

- from the main processor to the coprocessor (command including addresses a and b and a *start* order),
- from the main processor to the coprocessor data memory (unfiltered image) and from the coprocessor data memory to the main processor (filtered image),

Fig. 7.24 Multi-master bus system

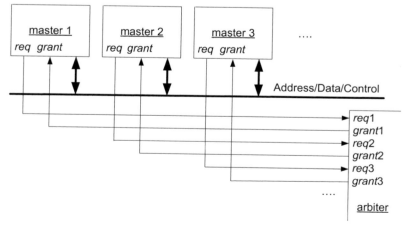

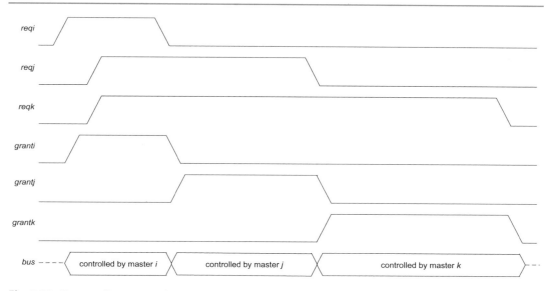

Fig. 7.25 Request–Grant protocol

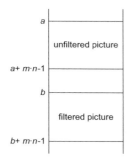

Fig. 7.26 Data memory

- from the coprocessor data memory to the coprocessor (reading of the unfiltered image) and from the coprocessor to its data memory (writing of the filtered image).

The proposed system configuration is shown in Fig. 7.27. A simple synchronous bus (Fig. 7.6) is used for all data transfers:

- In the case of data transmission from or to the main processor, the latter is the bus master.
- When the main processor is not involved in the data transmission, the coprocessor is the bus master.

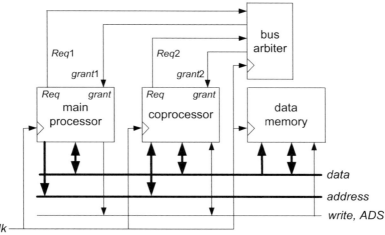

Fig. 7.27 Image processing system

In this way, the main processor can execute other internal operations while waiting for completion of the command execution. Thus, there are two bus masters, the main processor and the coprocessor, and an arbiter circuit is necessary.

The operations executed by the main processor are described by the following algorithm.

Algorithm 7.1 Program executed by the main processor

The operations executed by the coprocessor depend on the chosen image filtering algorithm. A very simple filter is used in this example. Consider a 3-by-3 mask $c(k, l)$ where $k, l \in \{-1, 0, 1\}$. Then, every pixel value is replaced by a linear combination of this pixel value and of the eight adjacent pixel values:

$$filtered\_pixel(i,j) = \sum_k \sum_l c(k,l) \cdot pixel(i+k,j+l). \quad (7.1)$$

```
--initial state: bus request signal inactive and
--outputs to the bus in high impedance
Req = 0; release(data, address, ADS, write);
--request bus control
Req = 1;
wait until Grant = 1;
--send unfiltered image to data memory
for i in 0 to m-1 loop
   for j in 0 to n-1 loop
      --write cycle (as bus master)
      data_memory(a + i·n + j) = pixel(i,j);
   end loop;
end loop;
--send addresses "a" and "b" and order
"start" to the
--coprocessor; three write cycles (as bus master)
coprocessor.a = a;
coprocessor.b = b;
coprocessor.start = true;
--release bus control: bus request signal inactive and
--outputs to the bus in high impedance
Req = 0; release(data, address, ADS, write);
--wait a few cycles while the arbiter grants the bus control --to the coprocessor;
wait for 100 cycles;
--request bus control; control will be granted when the
--coprocessor completes its operations
Req = 1;
wait until Grant = 1;
--read filtered image from the data memory
for i in 0 to m-1 loop
   for j in 0 to n-1 loop
      --read cycle (as bus master)
      filtered_pixel(i,j)= data_memory(b + i·n + j);
   end loop;
end loop;
```

p(0,0)	p(0,2)	p(0,3)	p(0,4)	p(0,5)	p(0,6)	p(0,7)	p(0,8)	p(0,9)	p(0,10)	p(0,11)	p(0,12)	p(0,13)	p(0,14)	p(0,15)
p(1,0)	p(1,2)	p(1,3)	p(1,4)	p(1,5)	p(1,6)	p(1,7)	p(1,8)	p(1,9)	p(1,10)	p(1,11)	p(1,12)	p(1,13)	p(1,14)	p(1,15)
P(2,0)	p(2,2)	p(2,3)	p(2,4)	p(2,5) c(0,0)	p(2,6) c(0,1)	p(2,7) c(0,2)	p(2,8)	p(2,9)	p(2,10)	p(2,11)	p(2,12)	p(2,13)	p(2,14)	p(2,15)
P(3,0)	P(3,2)	p(3,3)	p(3,4)	p(3,5) c(1,0)	p(3,6) c(1,1)	p(3,7) c(1,2)	p(3,8)	p(3,9)	p(3,10)	p(3,11)	p(3,12)	p(3,13)	p(3,14)	p(3,15)
p(4,0)	p(4,2)	p(4,3)	p(4,4)	p(4,5) c(2,0)	p(4,6) c(2,1)	p(4,7) c(2,2)	p(4,8)	p(4,9)	p(4,10)	p(4,11)	p(4,12)	p(4,13)	p(4,14)	p(4,15)
p(5,0)	p(5,2)	p(5,3)	p(5,4)	p(5,5)	p(5,6)	p(5,7)	p(5,8)	p(5,9) c(0,0)	p(5,10) c(0,1)	p(5,11) c(0,2)	p(5,12)	p(5,13)	p(5,14)	p(5,15)
p(6,0)	p(6,2)	p(6,3)	p(6,4)	p60,5)	p(6,6)	p(6,7)	p(6,8)	p(6,9) c(1,0)	p(6,10) c(1,1)	p(6,11) c(1,2)	p(6,12)	p(6,13)	p(6,14)	p(6,15)
p(7,0)	p(7,2)	p(7,3)	p(7,4)	p(7,5)	p(7,6)	p(7,7)	p(7,8)	p(7,9) c(2,0)	p(7,10) c(2,1)	P(7,11) c(2,2)	p(7,12)	p(7,13)	p(7,14)	p(7,15)
p(8,0)	p(8,2)	p(8,3)	p(8,4)	p(8,5)	p(8,6)	p(8,7)	p(8,8)	p(8,9)	p(8,10)	p(8,11)	p(8,12)	p(8,13)	p(8,14)	p(8,15)

Fig. 7.28 Image filtering

Consider the 9-by-16 image p of Fig. 7.28. As an example, pixel $p'(3,6)$ of the filtered image is equal to

$$p'(3,6) = c(0,0) \cdot p(2,5) + c(0,1) \cdot p(2,6)$$
$$+ c(0,2) \cdot p(2,7) + c(1,0) \cdot p(3,5)$$
$$+ c(1,1) \cdot p(3,6) + c(1,2) \cdot p(3,7)$$
$$+ c(2,0) \cdot p(4,5) + c(2,1) \cdot p(4,6)$$
$$+ c(2,2) \cdot p(4,7),$$

and pixel $p'(6,10)$ of the filtered image is equal to

$$p'(6,10) = c(0,0) \cdot p(5,9) + c(0,1) \cdot p(5,10)$$
$$+ c(0,2) \cdot p(5,11) + c(1,0) \cdot p(6,9)$$
$$+ c(1,1) \cdot p(6,10) + c(1,2) \cdot p(6,11)$$
$$+ c(2,0) \cdot p(7,9) + c(2,1) \cdot p(7,10)$$
$$+ c(2,2) \cdot p(7,11).$$

The operations executed by the coprocessor are described by the following algorithm.

Algorithm 7.2 Program executed by the coprocessor

```
--initial state: bus request signal inactive and
--outputs to the bus in high impedance
Req = 0; release(data, address, ADS, write);
--branch (command decoding)
if write_cycle and address_in = address.a then
   a = data_in;
elsif write_cycle and address_in = address.b then
   b = data_in;
elsif write_cycle and address_in = address.start then
   --request bus control
   Req = 1;
   wait until Grant = 1;
```

```
--filtering operations
for i in 1 to m-2 loop
   for j in 1 to m-2 loop
      acc = 0;
      for k in -1 to 1 loop
         for l in -1 to 1 loop
            --read data memory (as bus master)
            --and update acc
            acc = acc + c(k,l)·data_memory(a+(i+k)·n +j+l);
         end loop;
      end loop;
      --write data memory (as bus master)
      data_memory(b + i·n + j) = acc;
   end loop;
end loop;
--release bus control: bus request signal inactive and
--outputs to the bus in high impedance
Req = 0; release(data, address, ADS, write);
else nop;
end if;
```

In order to avoid negative pixel coordinates when computing (7.1), the filtered image has been reduced to m-2 rows and n-2 columns (left, right, up and down edges are deleted).

A data flow model of the complete system is available at the Authors' web site. It consists of five VHDL files: *system.vhd*, *cpu.vhd*, *arbiter.vhd*, *coprocessor.vhd* and *data_memory.vhd*. The parameter values and the original unfiltered image are defined within a package included in *coprocessor.vhd*. In particular, the addresses are 10-bit vectors defined as shown in Table 7.1.

Table 7.1 Memory map

Address	Addressed data
0000000000	*coprocessor.a*
0010000000	*coprocessor.b*
0100000000	*coprocessor.start*
1 a_8 a_7 a_6 a_5 a_4 a_3 a_2 a_1 a_0	*data_memory*(a_8 a_7 a_6 a_5 a_4 a_3 a_2 a_1 a_0)

Parts of the simulation results are shown in Figs. 7.29, 7.30 and 7.31. The unfiltered 9×16 image is

```
CONSTANT unfiltered: picture := (
1, 1, 2, 3, 4, 5, 6, 7, 8, 7, 6, 5, 4, 3, 2, 1,
1, 1, 2, 3, 4, 5, 6, 7, 8, 7, 6, 5, 4, 3, 2, 1,
4, 4, 4, 4, 4, 4, 4, 4, 4, 4, 4, 4, 4, 4, 4, 4,
1, 1, 2, 3, 4, 5, 6, 7, 8, 7, 6, 5, 4, 3, 2, 1,
1, 1, 2, 3, 4, 5, 6, 7, 8, 7, 6, 5, 4, 3, 2, 1,
4, 4, 4, 4, 4, 4, 4, 4, 4, 4, 4, 4, 4, 4, 4, 4,
1, 1, 2, 3, 4, 5, 6, 7, 8, 7, 6, 5, 4, 3, 2, 1,
1, 1, 2, 3, 4, 5, 6, 7, 8, 7, 6, 5, 4, 3, 2, 1,
4, 4, 4, 4, 4, 4, 4, 4, 4, 4, 4, 4, 4, 4, 4, 4
);
```

and the mask coefficients are

```
CONSTANT c: mask := (
x"ff", x"ff", x"ff",
x"ff", x"08", x"ff",
x"ff", x"ff", x"ff"
);
```

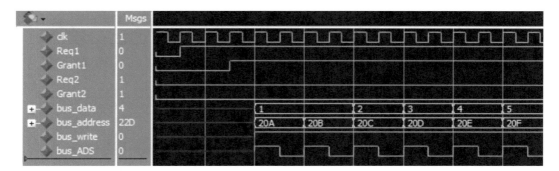

Fig. 7.29 Data memory writing (unfiltered picture) (courtesy of Mentor Graphics)

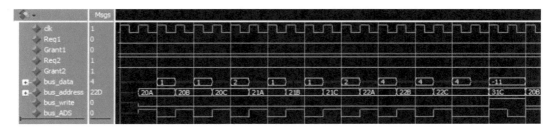

Fig. 7.30 First pixel processing (courtesy of Mentor Graphics)

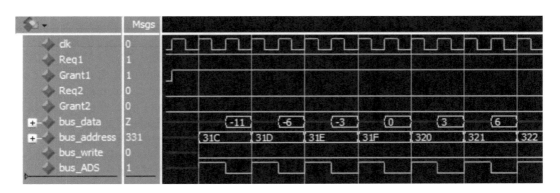

Fig. 7.31 Data memory reading (filtered picture) (courtesy of Mentor Graphics)

so that

$$filtered\_picture(1,1) = (-1) \cdot 1 + (-1) \cdot 1$$
$$+ (-1) \cdot 2 + (-1) \cdot 1$$
$$+ 8 \cdot 1 + (-1) \cdot 2$$
$$+ (-1) \cdot 4 + (-1) \cdot 4$$
$$+ (-1) \cdot 4 = -11,$$

$$filtered\_picture(1,2) = (-1) \cdot 1 + (-1) \cdot 2$$
$$+ (-1) \cdot 3 + (-1) \cdot 1$$
$$+ 8 \cdot 2 + (-1) \cdot 3$$
$$+ (-1) \cdot 4 + (-1) \cdot 4$$
$$+ (-1) \cdot 4 = -6,$$

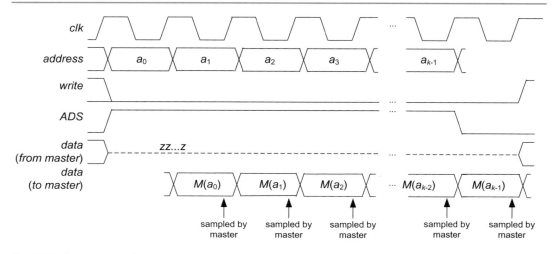

Fig. 7.32 Successive read operations

$$filtered\_picture(1, 3) = (-1) \cdot 2 + (-1) \cdot 3$$
$$+ (-1) \cdot 4 + (-1) \cdot 2$$
$$+ 8 \cdot 3 + (-1) \cdot 4$$
$$+ (-1) \cdot 4 + (-1) \cdot 4$$
$$+ (-1) \cdot 4 = -3,$$

and so on. Figure 7.29 shows the first steps of the transmission to the coprocessor data memory of the unfiltered picture, Fig. 7.30 shows the first step of the filtering operation, and Fig. 7.31 shows the first steps of the transmission to the main processor of the filtered picture.

To conclude this section on buses, three comments.

• Many other bus features and configurations can be defined. As an example, the image processing system (Example 7.4) suggests an improved communication protocol: when the main processor reads pixel values from the data memory, $m \cdot n$ successive 2-cycle read operations are executed (Fig. 7.31). A unique k-cycle ($k = m \cdot n + 1$) read operation could

also be considered (Fig. 7.32), so that the complete transfer time would be equal to $m \cdot n + 1$ cycles instead of $2 \cdot m \cdot n$. Similarly instead of $m \cdot n$ successive 2-cycle write operations, a unique k-cycle write operation could be considered.

• Another comment similar to the previous: in the case of systems that process large-width data (structured data, arrays), it can be useful to add the possibility of burst transmission. If n-bit data must be transmitted, where $n = s \cdot m$, then in order to reduce the number of bits of the input and output ports every data can be transmitted as a burst of s successive m-bit number. An example of synchronous bus, with burst read and write capabilities, is shown in Fig. 7.33. The master component defines a base address (a) and a data size s (4 in the example). Then m-bit words are read from or written to the selected slave at addresses a to $a + size$-1.

• Modern IC and FPGA electronic design automation packages include tools that permit to synthesize input and output interfaces for several predefined standard and complex buses.

Fig. 7.33 Synchronous bus with burst read (**a**) and burst write (**b**) capabilities

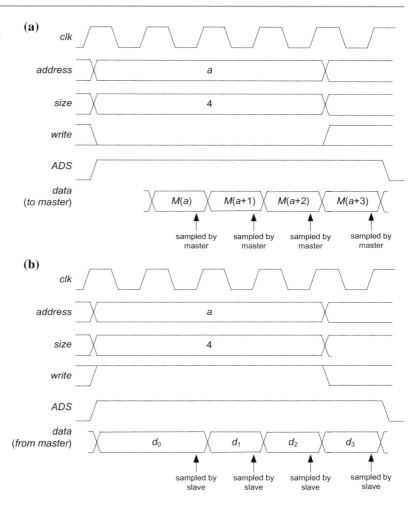

7.3 Exercises

1. Design an interface (bridge) between a synchronous bus with waiting cycles and an asynchronous bus.
2. Design a new version of the image processing system of Example 7.4 with two buses: a synchronous bus with waiting cycles to which are connected the main processor, the data memory and the bus bridge and an asynchronous bus to which are connected the coprocessor and the bus bridge. Generate a VHDL model of the complete system.

Bibliography

Deschamps JP, Valderrama E, Terés Ll (2017) Digital Systems: from Logic Gates to Processors. Springer, New York

Development Tools 8

The implementation of a digital circuit in an application-specific integrated circuit (ASIC), in a field-programmable gate array (FPGA) or in a complex programmable logic device (CPLD) is divided into several steps. The first step is the specification of the circuit. It includes a functional description, for example, an algorithm that the circuit must execute, as well as other characteristics such as maximum computation time, maximum power consumption, maximum chip size (ASIC), or maximum number of cells (FPGA, CPLD). If the functional specification is an algorithm, it can be defined in a very precise form by a program in some languages such as C/C++, Java, SystemC, MATLAB, or even in a hardware description language (HDL) at functional level. The resulting code can be executed so that the designer has the possibility to check whether the initial specification accurately defines the function that he actually wants to implement.

Once the functional specification has been approved, the next step is the generation of a digital circuit made up of available components such as logic gates, registers, and memory blocks. Nowadays several electronic design automation (EDA) tools help the designer to translate the initial specification to a logic circuit and to check the correctness of the so-obtained circuit.

The remaining steps depend on the targeted implementation technology. According to criteria such as cost, frequency, time to market, and so on, a Standard Cell (SC), Gate Array (GA), FPGA or CPLD family is selected and the previously obtained logic circuit is redesigned using components of the corresponding SC, GA, FPGA or CPLD libraries.

In this chapter, the main concepts related to the EDA tools are briefly described. Examples using Vivado HLS (Xilinx) and ISE (Xilinx) tools are given.

8.1 Design Flow

The design flows are the combination of EDA tools to carry out a circuit design. Figure 8.1 shows an example of design flow. It starts from an initial specification in a programming language or even in a hardware description language at functional level. The resulting code can be executed to check whether the initial specification accurately defines the function to be implemented.

The translation of this initial specification to a digital circuit, made up of available components, is decomposed into two steps. First a register-transfer level (RTL) description is generated: It defines the working of the circuit cycle by cycle; that means that the schedule of the operations is explicitly defined. This RTL description is generally expressed in a standard HDL, so that RTL description is another possible design entry point. The RTL description can be simulated to check whether the obtained circuit model accurately defines the function to be implemented and fulfills the expected timing characteristics (e.g., number of clock cycles).

© Springer Nature Switzerland AG 2019
J.-P. Deschamps et al., *Complex Digital Circuits*,
https://doi.org/10.1007/978-3-030-12653-7_8

Fig. 8.1 Example of design flow

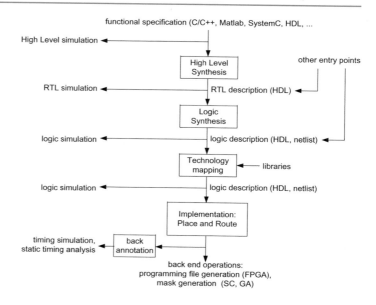

The second part of the design work is the translation from a RTL definition to a circuit made up of available components such as logic gates, registers, and memory blocks. In Fig. 8.1, the translation to a logic circuit is divided into two steps. First, a circuit made up of generic gates, registers, memory blocks, and so on, is defined. Then the obtained logic circuit is mapped to an equivalent logic circuit made up of elements of the chosen implementation library (Standard Cell, Gate Arrays, FPGA, CPLD libraries). Those logic descriptions can be expressed in several ways, for example a standard HDL, a standard netlist language such as EDIF (Electronic Design Interchange Format), a vendor-specific netlist language. Thus, logic description is another possible design entry point. Graphical user interfaces can also be used (e.g., schematic capture). The logic descriptions can be simulated to check whether the obtained circuit models accurately define the function to be implemented and whether they fulfill the expected timing characteristics and reach the expected cost and size limits (e.g., number of gates).

Once a logic circuit made up of components of the chosen implementation libraries has been defined, the implementation step is performed. Those library elements are placed and routed within a particular component with its

corresponding package. The implemented circuit description is used to generate the information necessary to execute the remaining (back end) operations: generation of programming files (FPGA, CPLD) or manufacturing of masks (SC, GA). Furthermore, accurate timing information predictions can be extracted (back annotation) from the implemented circuit description. This information can be used to simulate the circuit with this timing information or to detect and analyze critical paths.

8.2   Logic Synthesis

Logic synthesis is a basic function of any EDA package. Its function is the translation of register-transfer level (RTL) descriptions to logic descriptions:

- RTL descriptions define the circuit working cycle by cycle, so that they explicitly define the schedule of the operations; many examples of RTL descriptions in VHDL have been presented in the preceding chapters.
- Logic descriptions are constituted of a list of components (gates, multiplexers, flip-flops, and so on) and of interconnections nets, a so-called *netlist*.

Logic synthesis can be divided into two steps:

- First (logic synthesis in "strictly speaking" sense) translation of the initial RTL description to logic descriptions consisting of generic gates, flip-flops, registers, and so on, without specification of the target technology.
- Then (technology mapping) mapping of the obtained logic circuit to an equivalent circuit made up of elements of the chosen implementation library (Standard Cell, Gate Arrays, FPGA, CPLD libraries).

In some cases, both steps are integrated within a unique synthesis tool. As an example, it is the case of development tools supplied by FPGA or CPLD vendors: The logic synthesis program directly generates netlists whose elements belong to the corresponding FPGA or CPLD basic component libraries.

The input to a logic synthesizer is an RTL description, for example, in some standard HDL.

The output is a structural description, for example in a standard HDL, a standard netlist language or even a proprietary (vendor specific) netlist format, as well as various reports (number of logic components, expected dynamic and electrical characteristics).

The synthesis process executes optimization tasks such as area reduction, speed optimization, and reduction of the power consumption. Furthermore, some aspects of the synthesis work can be controlled by the designer. For example, the way the internal states of a finite-state machine are encoded (minimum number of state variables, one variable per state, and so on), the maximum fan-out of nets and other constraints, the use of predefined components such as memory or arithmetic blocks, and others.

Example 8.1 The following VHDL code (*counter.vhd* available at the Authors' web site) defines a mod 8 counter, with count enable input, at RT level:

```
LIBRARY...
ENTITY counter IS
PORT (
   clk, reset, count_enable: IN STD_LOGIC;
   y: OUT STD_LOGIC_VECTOR(2 DOWNTO 0));
END counter;

ARCHITECTURE behavior OF counter IS
   SIGNAL next_state, current_state: STD_LOGIC_VECTOR(2 DOWNTO 0);
BEGIN
   next_state <= current_state + count_enable;
   synchronization: PROCESS(reset, clk)
   BEGIN
      IF reset = '1' THEN current_state <= "000";
      ELSIF clk'EVENT AND clk = '1' THEN current_state <= next_state;
   END IF;
   END PROCESS synchronization;
    y <= current_state;
end behavior;
```

On every positive edge of *clk*, the process *synchronization* replaces the value of *current_state* by *next_state = current_state + count_enable*, so that if *count_enable* = 1 then the updated state is equal to *current_state* + 1, and if *count_enable* = 0 then the updated state is equal to *current_state* (the internal state does not change). The three-bit output signal *y* is equal to *current_state*. The logic synthesis tool XST of ISE (Xilinx) generates the circuit of Fig. 8.2 consisting of three input buffers (*clk*, *reset*, *count_enable*), three output buffers (*y*), three flip-flops and two lookup tables, and an inverter that generates the following functions

$$result_0 = not(cs_0), \tag{8.1a}$$

$$result_1 = cs_0 \oplus cs_1, \tag{8.1b}$$

$$result_2 = cs_0 \cdot cs_1 \oplus cs_2. \tag{8.1c}$$

Thus, if *count_enable* = 0, then the three flip-flop *CE* (clock enable) inputs are equal to 0 so that the internal state *cs* (*current_state*) does not change. When *count_enable* = 1 then, according to Eq. (8.1) that define a 3-bit half adder (*result* = *cs* +1 mod 8), the internal state *cs* is replaced by *cs* + 1 mod 8.

The VHDL file *counter_synthesis.vhd*, automatically generated by XST, describes the circuit of Fig. 8.2. It is available at the Authors' web site. It uses the following components of UNI-SIM (the library for functional simulation of Xilinx primitives): two input buffers IBUF, a clock input buffer BUFGP, three output buffers OBUF, three FDCE flip-flop, an inverter INV, a 2-input lookup table LUT2 whose contents are defined by a generic parameter equal to "6" (Table 8.1) and a 3-input lookup table LUT3 whose contents are defined by a generic parameter equal to "6a" (Table 8.2).

Table 8.1 LUT2 with generic parameter equal to "6" = "0110": O = $I_1 \oplus I_0$

I_1	I_0	O
0	0	0
0	1	1
1	0	1
1	1	0

Table 8.2 LUT3 with generic parameter equal to "6A" = "01101010": O = $I_2 \cdot I_1 \oplus I_0$

I_2	I_1	I_0	O
0	0	0	0
0	0	1	1
0	1	0	0
0	1	1	1
1	0	0	0
1	0	1	1
1	1	0	1
1	1	1	0

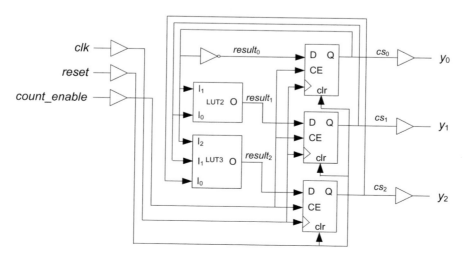

Fig. 8.2 Mod 8 counter

After the synthesis of the counter with generic logic components, the next step (technology mapping) is the generation of an equivalent logic circuit made up of logic component belonging to the specific libraries corresponding to the chosen device. It is an intermediate and necessary step between logic synthesis and implementation.

8.3   High-Level Synthesis

Commercial high-level synthesis (HLS) tools are now available. Their function (Fig. 8.1) is the translation of a high-level description to an RTL description. The input to a high-level synthesizer

possibility to guide the HLS tool by inserting directives (pragmas) within the initial description.

Example 8.2 The following C code defines a circuit that computes $a \cdot 5^{10}$.

```
unsignedinta_loop(unsignedint a)
{
    for (int i = 0; i < 10; i ++)a = 5*a;
    return a;
}
```

The following "finite-state machine with operations" (Sect. 6.2 of Deschamps et al. 2017) is an RTL description of the corresponding circuit:

```
fsm_with_operations: process(clk)
begin
if (clk'event and clk = '1') then
 if (rst = '1') then CS <= "0"; done <= '0';
 elsif (CS = "0") and (start = '1') then
     CS <= "1"; int_a <= a; i <= "0000"; done <= '0';
   elsif (CS = "1") and (exit_cond = "0") then
     int_a <= shl(int_a, two) + int_a;
     i <= i + 1; done <= '0';
   elsif (CS = "1") and (exit_cond = "1") then
     CS <= "0"; done <= '1';
   end if;
end if;
end process;
data_return <= int_a;
exit_cond <= "1" when (i = "1010") else "0";
```

tool is the description of the function executed by the circuit in a language such as C/C++, Java, SystemC, MATLAB (especially for digital signal processing functions), and others. The output is a RTL description, generally in a standard HDL, as well as various reports (e.g., number of clock cycles). Thus, HLS tools must be able to execute the tasks described in the preceding chapters, for example, scheduling, resource assignment, pipelining, loop unrolling, generation of input–output interfaces, and so on. The designer has the

It is a 2-state finite-state machine with operations whose current internal state *CS* and condition signal (flag) *exit_cond* has been defined as 1-bit vectors. It works as follows:

- initially (on reset) $CS = 0$ and *done* = 0;
- when $CS = 0$: if *start* = 1 then input *a* is internally stored within *int_a*, index *i* is set to 0, $CS = 1$ and *done* = 0;
- when $CS = 1$: as long as *exit_cond* is false (= 0), *int_a* is replaced by $5 \cdot int\_a = 4 \cdot$

int_a + *int_a* = (*int_a* shifted two bits to the left) + *int_a*, and *i* is replaced by *i* + 1;
- when *CS* = 1 and *exit_cond* is true (= 1) then *CS* = 0 and *done* = 1.

The *exit_cond* signal is equal to 1 if and only if *i* is equal to 10, and the output *data_return* is equal to *int_a*. The VHDL file *rtl_loop.vhd* including the preceding RTL description is available at the Authors' web site. The VHDL file *a_loop.vhd* automatically generated by Vivado HLS (Xilinx) is also available at the Authors' web site, with some modification in what respects the identifier names, and is equivalent to the VHDL file *rtl_loop.vhd*. It includes the definition of two additional output signals *ready* and *idle* that can be used to control the communication with other circuits (handshake protocols).

Commercial HLS tools give the designer the possibility to guide the synthesis process by inserting directives (pragmas) within the initial description.

Example 8.3 The following C code defines a circuit that computes the same function as the code of Example 8.2 with an additional pragma "unroll":

```
unsigned int a_loop(unsigned int a)
{
        for (int i = 0; i < 10; i ++)
#pragma AP unroll
        a = 5*a;
        return a;
}
```

The following sequence of signal assignment statements defines an iterative combinational circuit, consisting of cells that compute $a_i = 4 \cdot a_{i-1} + a_{i-1} = 5 \cdot a_{i-1}$, $i = 1$–9, with $a_0 = a$, so that the final result is $a \cdot 5^{10}$:

```
a_1  <= shl(a, two)   + a;
a_2  <= shl(a_1, two) + a_1;
a_3  <= shl(a_2, two) + a_2;
a_4  <= shl(a_3, two) + a_3;
```

```
a_5  <= shl(a_4, two) + a_4;
a_6  <= shl(a_5, two) + a_5;
a_7  <= shl(a_6, two) + a_6;
a_8  <= shl(a_7, two) + a_7;
a_9  <= shl(a_8, two) + a_8;
data_return <= shl(a_9, two) + a_9;
```

A complete VHDL file *unroll_loop.vhd* is available at the Authors' web site. It is equivalent to the VHDL file automatically generated by Vivado HLS (Xilinx).

Example 8.4 The following C code defines a circuit that computes the same function as in the preceding Examples 8.2 and 8.3 with an additional pragma specifying a pipeline circuit with an input interval (II) equal to 1 clock period:

```
unsigned int a_loop(unsigned int a)
{
#pragma HLS PIPELINE II = 1
        for (int i = 0; i < 10; i ++)
        a = 5*a;
        return a;
}
```

The synthesized circuit is similar to that of Example 3 with additional pipeline registers inserted between successive iterative cells. The following sequence of signal assignment statements defines ten combinational cells that compute $a_i = 4 \cdot a_{i-1d} + a_{i-1d} = 5 \cdot a_{i-1d}$, $i = 1$–9, with $a_{0d} = a$ and where $a_{i\,d}$ is equal to signal a_i delayed one cycle. Thus, the final result is equal to $a \cdot 5^{10}$ with a total delay equal to nine clock cycles:

```
a_1  <= shl(a, two)   + a;
a_2  <= shl(a_1d, two) + a_1d;
a_3  <= shl(a_2d, two) + a_2d;
a_4  <= shl(a_3d, two) + a_3d;
a_5  <= shl(a_4d, two) + a_4d;
a_6  <= shl(a_5d, two) + a_5d;
a_7  <= shl(a_6d, two) + a_6d;
a_8  <= shl(a_7d, two) + a_7d;
a_9  <= shl(a_8d, two) + a_8d;
data_return <= shl(a_9d, two) + a_9d;
```

The complete circuit includes nine pipeline registers that generate $a_{id}(n \cdot T) = a_i((n-1) \cdot T)$, $i = 1–9$, being T the clock period:

```
reg_i: process(clk)
begin
   if clk'event and clk = '1' then
      if it(i-1) = '1' then a_id <= a_i; end if;
   end if;
end process;
```

The pipeline register enable signals $it(0–9)$ are generated by a shift register whose serial input is *start* and whose serial output is *done*:

```
it0 <= start;
shift_register: process(clk)
begin
   if clk'event and clk = '1' then
      if rst = '1' then it <= (others => '0');
      else it <= it0&it(1 to 8);--right shift
      end if;
   end if;
end process;
done <= it(9);
```

A complete VHDL file *pp_loop.vhd* is available at the Authors' web site. It is equivalent to the VHDL file automatically generated by Vivado HLS (Xilinx).

In fact, many of the operations described in the preceding chapters can be executed by commercial HLS tools. Some examples of circuit characteristics that the designer can define by using high-level language constructs, declarations, directives, are the following:

- definition of specific data types, for example, binary vectors with non-standard bit width, array partitioning, array merging;
- operation scheduling (Chap. 2), for example, minimum and maximum latency, clock frequency;
- resource assignment (Chap. 2), for example, use of virtual components (IP cores) belonging to available libraries;

- pipelining with predefined introduction interval definition, latency, throughput (Chap. 3, Example 8.4);

- loop implementation, for example, loop unrolling (Chap. 4, Example 8.3), loop flattening (collapsing of nested loops into a single loop), merging of consecutive loops;

- data path optimization techniques, for example use of buses, transmission of data through FIFO memories (streaming), inlining (Chap. 5);
- memory implementation options, for example, blocks of RAM, FIFO, FILO (Chap. 5);
- interfaces, for example, handshaking protocols (Chap. 7).

8.4 Implementation

The implementation step is executed by a vendor-specific tool that places and routes the design in the target device. The inputs to the implementation tool are the previously synthesized files as well as implementation constraints, mainly timing and area constraints. The output is a proprietary placed and routed netlist.

The implementation tools generate information about the implemented circuit. For example

- detailed information about the used resources,
- information about the clock distribution network,
- final timing compared with the implementation constraints,
- delay of internal interconnections,
- graphical view of the placed and routed design within the target device,
- power consumption estimation.

The placed and routed netlist is used to generate the information necessary to program or to manufacture the circuit: programming files (FPGA, CPLD) or sets of masks (SC, GA).

8.5   Simulation

Commercial simulation tools are available for every type of circuit definition, from functional (or behavioral) level to transistor level. In Sect. 8.2, two types of design tools have been described: logic synthesis (from register-transfer level to logic components level), and high-level synthesis (from functional level to register-transfer level). So, as regards simulation tools, three description levels should be considered: functional, register transfer and logic components levels.

Whatever the description level, simulation tools—apart from the simulation program itself—include the following designer interfaces:

- an input language or a visual interface to define the circuit to be simulated;
- a language or a visual interface to define the values of the input signals during each simulation step (stimuli, test vectors);
- a textual or visual interface to observe the simulation results.

Consider the three types of simulation tools mentioned above.

8.5.1   Functional Simulation

HLS descriptions define the circuit function in languages such as C/C++, SystemC, or even HDL's at behavioral level. The language used to specify the circuit function can also be used to define the values of the input signals.

Example 8.5 Consider again the functional definition of Example 8.2. The following C code defines a circuit that computes $a \cdot 5^{10}$.

```
unsignedinta_loop(unsignedint a)
{
    for (int i = 0; i < 10; i ++)a = 5*a;
    return a;
}
```

To check the correctness of this description, a C program that generates successive values of a, and for each of them executes a call to function a_loop, is defined (a so-called test bench):

```
#include < iostream>
using std::cout;
using std::cin;
unsignedinta_loop(unsignedint a);
int main()
{
    unsignedint a[10] = {2, 7, 5, 4, 3, 9, 1, 0, 9, 8};
    unsignedint result[10];
    for (int i = 0; i < 10; i ++)
    {
        result[i] = a_loop(a[i]);
        cout < < a[i]; cout < < "·(5 exp 10) = ";
        cout < < result[i]; cout < < std::endl;
    }
}
```

Successive values of *a* are defined (2, 7, 5, ...) and the "under test" function *a_loop* is executed with those successive values of *a*. In this elementary example, the standard user interfaces (keyboard, console) are used. The result is the following:

```
2·(5 exp 10) = 19531250
7·(5 exp 10) = 68359375
5·(5 exp 10) = 48828125
4·(5 exp 10) = 39062500
3·(5 exp 10) = 29296875
9·(5 exp 10) = 87890625
1·(5 exp 10) = 9765625
0·(5 exp 10) = 0
9·(5 exp 10) = 87890625
8·(5 exp 10) = 78125000
```

file *ap_int.h* available within Vivado HLS (Xilinx).

```
#include "ap_int.h"
typedef ap_int < 8>     data;
typedef ap_uint < 4>    instruction_field;
typedef ap_uint < 8>    address;
typedef ap_uint < 16 >  instruction;
```

The circuit processes 2's complement 8-bit numbers whose type is called *data*. It executes 16-bit *instructions* partitioned into four 4-bit *fields*. The instruction memory stores 256 16-bit words and has 8-bit *addresses*. The preceding definitions are saved within a file *hls_processor.h*. The processor is specified as a C function.

```
#include "hls_processor.h"
voidhls_processor (
bool reset, data IN0, data IN1, data IN2, data IN3, data IN4,
data IN5, data IN6, data IN7, instruction_field code,
instruction_field f1, instruction_field f2, instruction_field f3,
data * OUT0, data * OUT1, data * OUT2, data * OUT3, data * OUT4, data * OUT5, data * OUT6,
data * OUT7, address * number)
```

In the preceding example, simulation steps correspond to successive values of the input signal *a*.

Example 8.6 Consider a second example in which the potentiality of a programming language description is much more obvious. In Chap. 5 of Deschamps et al. (2017), a simple 8-bit microprocessor is developed, starting from a functional definition. The initial functional specification can be described in C. First, several specific data types are defined using a predefined

The input and output variables are

- *reset* (binary),
- IN_0 to IN_7: the eight 8-bit input ports,
- *code*, f_1, f_2 and f_3: the four 4-bit instruction fields,
- OUT_0 to OUT_7: the eight 8-bit output ports,
- *number*: 8-bit instruction memory address.

Three internally registered variables and an 8-element array are declared at the beginning of the function description:

```
{
static data X[16];
static address pc;
static data output_port [8];
data input_port[8];
input_port[0] = IN0; input_port[1] = IN1; input_port[2] = IN2;
input_port[3] = IN3; input_port[4] = IN4; input_port[5] = IN5;
input_port[6] = IN6; input_port[7] = IN7;
```

- *X* is a 16-word 8-bit register file, *pc* is the 8-bit program counter and *output_port* is the set of eight 8-bit output registers; they are internal registers of the processor and must be modeled by static variables that maintain their value between successive function calls (global variables);

- *input_port* is the set of the eight 8-bit input ports.

The function body definition is the following:

```
if (reset == 1) pc = 0;
else {
      switch (code)
      {
            case 0: {
                  X[f3] = f2 + 16*f1; pc = pc + 1;
                  }
                  break;
            case 2: {
                  X[f3] = input_port[f3]; pc = pc + 1;
                  }
                  break;
            case 10: {
                  output_port[f1] = X[f2]; pc = pc + 1;
                  }
                  break;
            case 8: {
                  output_port[f1] = f3 + 16*f2; pc = pc + 1;
                  }
                  break;
            case 4: {
                  X[f3] = X[f1] + X[f2]; pc = pc + 1;
                  }
                  break;
            case 5: {
                  X[f3] = X[f1] - X[f2]; pc = pc + 1;
                  }
                  break;
            case 14: {
                  pc = f3 + 16*f2;
                  }
                  break;
            case 12: {
                  if (X[f1] > 0) pc = f3 + 16*f2;
                  else pc = pc + 1;
                  }
                  break;
            case 13: {
                  if (X[f1] < 0) pc = f3 + 16*f2;
                  else pc = pc + 1;
                  }
                  break;
      }
}
```

```
*OUT0 = output_port[0]; *OUT1 = output_port[1];
*OUT2 = output_port[2]; *OUT3 = output_port[3];
*OUT4 = output_port[4]; *OUT5 = output_port[5];
*OUT6 = output_port[6]; *OUT7 = output_port[7];
*number = pc;}
```

Initially, the program counter is set to 0. Then the instruction code field is read, the corresponding operation is executed and the program

To check the correctness of this description, a test bench must be generated. For that a new heading file *temp_control.h* is defined.

```
#include "hls_processor.h"
const instruction_field ccode [16] = {0x0, 0x2, 0x2, 0x5, 0xd, 0xc, 0xe, 0x8, 0xe, 0x8,
0x2, 0x2, 0x5, 0x5, 0xd, 0xe};
const instruction_field ii [16] =   {0x0, 0x0, 0x0, 0x0, 0x4, 0x4, 0x0, 0x0, 0x0, 0x0,
0x0, 0x0, 0x2, 0x4, 0x4, 0x0};
const instruction_field jj [16] =   {0xA, 0x0, 0x1, 0x1, 0x0, 0x0, 0x0, 0x0, 0x0, 0x0,
0x2, 0x2, 0x3, 0x5, 0x0, 0x0};
const instruction_field kk [16] =   {0x5, 0x0, 0x1, 0x4, 0x7, 0x9, 0xA, 0x1, 0xA, 0x0,
0x3, 0x2, 0x4, 0x4, 0xB, 0x1};
```

counter is updated. The instruction encoding and the corresponding operations are defined Table 8.3 where A is an 8-bit immediate value, N is an 8-bit program memory address, and i, j and k are 4-bit vectors that address the 16-word internal register file X.

It defines the instruction memory contents that correspond to a temperature control program executed on the processor. This program consists of sixteen instructions and is described in (Deschamps et al. 2017, Fig. 5.3). The four instruction fields are defined separately. The test bench starts with head file inclusions, interface definitions, and declaration of the function under test (*hls_processor*):

```
#include "hls_processor.h"
#include "temp_control.h"
#include < stdio.h>
#include < iostream>
using std::cout;
using std::cin;
voidhls_processor (bool reset, data IN0, data IN1, data IN2, data IN3, data IN4, data
IN5, data IN6, data IN7, instruction_field code, instruction_field f1, instruction_field
f2, instruction_field f3,
data * OUT0, data * OUT1, data * OUT2, data * OUT3, data * OUT4,
data * OUT5, data * OUT6, data * OUT7, address * number);
```

Table 8.3 Instruction encoding

Mnemonic	Code	f_1	f_2	f_3	Operation
assign_value	0	$A_{7.4}$	$A_{3.0}$	k	$X_k = A$
data_input	2	–	j	k	$X_k = IN_j$
data_output	10	i	j	–	$OUT_i = X_j$
output_value	8	i	$A_{7.4}$	$A_{3.0}$	$OUT_i = A$
operation_add	4	i	j	k	$X_k = X_i + X_j$
operation_sub	5	i	j	k	$X_k = X_i - X_j$
jump	14	–	$N_{7.4}$	$N_{3.0}$	go to N
jump_pos	12	i	$N_{7.4}$	$N_{3.0}$	if $X_i > 0$ go to N
jump_neg	13	i	$N_{7.4}$	$N_{3.0}$	if $X_i < 0$ go to N

The main function starts with the declaration of the test bench outputs and inputs:

- *onoff* is the circuit output connected to output port 0, and $port_1$ to $port_7$ correspond to the other output ports (not used in this temperature control application),
- *temp* (current temperature), *pos* (reference temperature) and *time* (current time) are

```
int main () {
data onoff, port1, port2, port3, port4,
port5, port6, port7;
ap_uint < 8> temp, time, pos;
address mem_addr;
instruction_field code, f1, f2, f3;
```

During the first simulation step the *hls_processor* function is executed with *reset* = 1, *temp* = 15, *pos* = 20 and *time* = 0:

```
temp = 15;
pos = 20;
time = 0;
hls_processor (1, temp, pos, time, 0, 0, 0, 0, 0, code, f1, f2, f3,
&onoff, &port1, &port2, &port3, &port4, &port5, &port6, &port7,
&mem_addr);
```

8-bit naturals connected to input ports 0, 1 and 2,

- *mem_address* is the instruction memory address connected to the processor output *number*,
- *code*, f_1, f_2 and f_3 are the 4-bit instructions fields read from the instruction memory.

Then, 51 simulation steps are executed with *reset* = 0, *temp* = 15 and *pos* = 20. At each step, the value of *time* is incremented. The values of *code*, f_1, f_2 and f_3 are read from the instructions memory at address *mem_address* and the *hls_processor* function is executed. The values of *time*, *pos*, *temp* and *onoff* are sent to the standard user interface.

```
for (int s = 0; s <=50; s ++) {
        time = time + 1;
        code = ccode[mem_addr]; f1 = ii[mem_addr]; f2 = jj[mem_addr];
        f3 = kk[mem_addr];
        hls_processor (0, temp, pos, time, 0, 0, 0, 0, 0,
        code, f1, f2, f3, &onoff, &port1, &port2, &port3,
        &port4,&port5, &port6, &port7,&mem_addr);
        cout << "time = "; cout << time; cout << " pos = ";
        cout << pos; cout << " temp = "; cout << temp;
        cout << " onoff = "; cout << onoff; cout << std::endl;
            }
temp = 25;
.........
temp = 18;
.........
}
```

The same operations are executed with other values of the current temperature (*temp* = 25 and *temp* = 18). As the reference temperature is always equal to 20, when the current temperature *temp* is equal to 15, *onoff* must be equal to 1; when *temp* = 25, *onoff* must be equal to 0; when *temp* = 18, *onoff* must be equal to 1. The simulation results are the following:

```
time = 1    temp = 15  mem_addr = 1   code = 0   onoff = 0
time = 2    temp = 15  mem_addr = 2   code = 2   onoff = 0
time = 3    temp = 15  mem_addr = 3   code = 2   onoff = 0
time = 4    temp = 15  mem_addr = 4   code = 5   onoff = 0
time = 5    temp = 15  mem_addr = 7   code = 13  onoff = 0
time = 6    temp = 15  mem_addr = 8   code = 8   onoff = 1
time = 7    temp = 15  mem_addr = 10  code = 14  onoff = 1
...
time = 51   temp = 15  mem_addr = 14  code = 5   onoff = 1
time = 52   temp = 25  mem_addr = 15  code = 13  onoff = 1
time = 53   temp = 25  mem_addr = 1   code = 14  onoff = 1
time = 54   temp = 25  mem_addr = 2   code = 2   onoff = 1
time = 55   temp = 25  mem_addr = 3   code = 2   onoff = 1
time = 56   temp = 25  mem_addr = 4   code = 5   onoff = 1
time = 57   temp = 25  mem_addr = 5   code = 13  onoff = 1
time = 58   temp = 25  mem_addr = 9   code = 12  onoff = 1
time = 59   temp = 25  mem_addr = 10  code = 8   onoff = 0
time = 60   temp = 25  mem_addr = 11  code = 2   onoff = 0
...
time = 102  temp = 25  mem_addr = 2   code = 2   onoff = 0
time = 103  temp = 18  mem_addr = 3   code = 2   onoff = 0
time = 104  temp = 18  mem_addr = 4   code = 5   onoff = 0
time = 105  temp = 18  mem_addr = 5   code = 13  onoff = 0
time = 106  temp = 18  mem_addr = 9   code = 12  onoff = 0
time = 107  temp = 18  mem_addr = 10  code = 8   onoff = 0
time = 108  temp = 18  mem_addr = 11  code = 2   onoff = 0
time = 109  temp = 18  mem_addr = 12  code = 2   onoff = 0
time = 110  temp = 18  mem_addr = 13  code = 5   onoff = 0
```

```
time = 111  temp = 18  mem_addr = 14 code = 5   onoff = 0
time = 112  temp = 18  mem_addr = 15 code = 13  onoff = 0
time = 113  temp = 18  mem_addr = 1  code = 14  onoff = 0
time = 114  temp = 18  mem_addr = 2  code = 2   onoff = 0
time = 115  temp = 18  mem_addr = 3  code = 2   onoff = 0
time = 116  temp = 18  mem_addr = 4  code = 5   onoff = 0
time = 117  temp = 18  mem_addr = 7  code = 13  onoff = 0
time = 118  temp = 18  mem_addr = 8  code = 8   onoff = 1
time = 119  temp = 18  mem_addr = 10 code = 14  onoff = 1
...
time = 153  temp = 18  mem_addr = 12 code = 2   onoff = 1
```

The response time between a current temperature change and the corresponding change of the *onoff* output depends on the current program counter value (*mem_addr*) when the temperature value changes. Complete C files *hls_processor.cc*, *hls_processor.h*, *temp_control.h* and *hls_processor_test.cc* are available at the Authors' web site.

8.5.2 Register-Transfer Simulation

RTL descriptions define the circuit working cycle by cycle, using for that hardware description languages. The same languages can also be used to generate test benches that define values of the input signals at each clock cycle.

Example 8.7 In Example 8.2, an initial functional specification (a C program) has been translated to an RTL specification *rtl_loop.vhd*. The following VHDL code (*test_rtl_loop.vhd* available at Authors' web site) simulates the circuit:

```
library ...
entity test_rtl_loop is end test_rtl_loop;
architecture test of test_rtl_loop is
   component rtl_loop is
   port (
      clk : IN STD_LOGIC;
      rst : IN STD_LOGIC;
      start : IN STD_LOGIC;
      done : OUT STD_LOGIC;
      a : IN STD_LOGIC_VECTOR (31 downto 0);
      data_return : OUT STD_LOGIC_VECTOR (31 downto 0));
   end component;
   signal clk: STD_LOGIC : = '1';
   signal rst, start, done: STD_LOGIC;
   signal a: STD_LOGIC_VECTOR (31 downto 0);
   signal data_return: STD_LOGIC_VECTOR (31 downto 0);
begin
   dut: rtl_loop port map (
      clk => clk, rst => rst, start => start, done => done, a => a,
      data_return => data_return);
   clk <= not(clk) after 50 ns;
   rst <= '1', '0' after 100 ns;
   start <= '0', '1' after 200 ns, '0' after 300 ns,
      '1' after 2500 ns, '0' after 2600 ns;
   a <= x"00000005", x"00000007" after 2500 ns;
end test;
```

The test bench *test_rtl_loop* instantiates the device under test (dut)—in this case the component *rtl_loop*—and defines the value of all input signals:

- *clk* is defined by an assignment that substitutes *clk* by *not(clk)* every 50 ns (a kind of oscillator), so that *clk* is a periodic signal with a period equal to 100 ns;
- *reset*, *start*, and *a* are explicitly defined in function of the current simulation time.

The compilation and execution of this test bench with a VHDL simulator permits to visualize the working of the circuit. Results corresponding to the computation of $5 \cdot 5^{10} = 48,828,125$ and $7 \cdot 5^{10} = 68,359,375$ are shown in Fig. 8.3.

A similar test bench of the pipelined implementation of the same function (Example 8.4) can be defined (*test_pp_loop.vhd* available at the Authors' web site). It first computes $5 \cdot 5^{10} = 48,828,125$, $7 \cdot 5^{10} = 68,359,375$ and $4 \cdot 5^{10} = 39,062,500$, with input interval equal to a clock period and delay equal to nine clock periods, under the control of a 3-cycle start pulse; it also computes $8 \cdot 5^{10} = 78,125,000$ under the control of a 1-cycle start pulse (Fig. 8.4).

In the preceding example, the input signal values are defined with simple signal assignments. In fact, a language such as VHDL permits to define circuits at several levels, not only at register-transfer level. So, when defining test benches, the whole potentiality of the language can be used.

Example 8.8 The following example is a radix 2^k adder (Sect. 7.3 of Deschamps et al. 2012). If $n = k \cdot m$, an n-bit adder can be implemented with m serially connected 2^k-bit adders (Fig. 8.5). This *carry-skip* adder structure has a shorter computation time than a simple *ripple carry* adder.

A complete VHDL model *base_2k_adder.vhd* is available at the Authors' web site. The following VHDL entity is an exhaustive test: all combinations of input signal values (x, y and c_{in}) are considered, where x and y are $k \cdot m$-bit numbers and c_{in} an initial carry. The output signal values are

$$z = (x + y + c_{in}) \bmod 2^n \text{ and } c_{out}$$
$$= (x + y + c_{in})/2^n. \tag{8.2}$$

Furthermore, the use of *assertion* statements permits to detect errors. For that, the value of $zz = x + y + c_{in}$ is computed and compared with the values of z and c_{out} generated by the circuit under test: if $z \neq zz \bmod 2^n$ or if $c_{out} \neq zz/2^n$ then

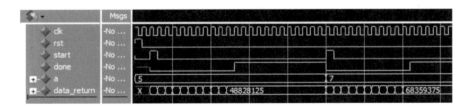

Fig. 8.3 Simulation results (*rtl_loop*) (courtesy of Mentor Graphics)

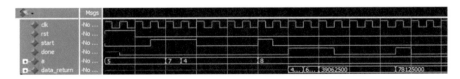

Fig. 8.4 Simulation results (*pp_loop*) (courtesy of Mentor Graphics)

Fig. 8.5 Radix 2^k adder cell

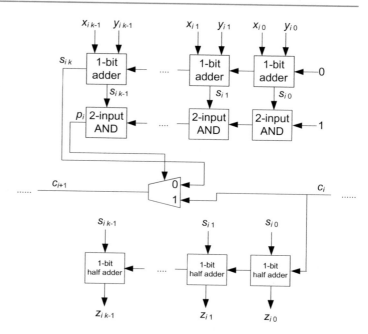

an error message is generated, including a short error description and the cycle during which it happens.

Consider a first test bench execution, introducing an error in the circuit description (*base_2k_adder.vhd*): the output carry c_{out} has been permanently connected to 0. The simulation

```
LIBRARY ...
ENTITY test_base_2k_adder IS END test_base_2k_adder;

ARCHITECTURE test OF test_base_2k_adder IS
    CONSTANT k: natural: = 2;
    CONSTANT m: natural: = 3;
    COMPONENT base_2k_adder IS
        GENERIC(k, m: NATURAL);
    PORT(
        x, y: IN STD_LOGIC_VECTOR(k*m-1 DOWNTO 0);
        c_in: IN STD_LOGIC;
        z: OUT STD_LOGIC_VECTOR(k*m-1 DOWNTO 0);
        c_out: OUT STD_LOGIC
    );
    END COMPONENT;
    SIGNAL x, y: STD_LOGIC_VECTOR(k*m-1 DOWNTO 0);
    SIGNAL c_in: STD_LOGIC;
    SIGNAL z: STD_LOGIC_VECTOR(k*m-1 DOWNTO 0);
    SIGNAL zz: STD_LOGIC_VECTOR(k*m DOWNTO 0);
    SIGNAL c_out: STD_LOGIC;
    CONSTANT DELAY : time : = 50 ns;
BEGIN
    dut: base_2k_adder GENERIC MAP(k => k, m => m)
    PORT MAP(x => x, y => y, c_in => c_in, z => z, c_out => c_out);
    zz <= ('0' & x) + y + c_in;
    stimuli: PROCESS
```

```
   BEGIN
   FOR i IN 0 TO 2**(k*m)-1 LOOP
       FOR j IN 0 TO 2**(k*m)-1 LOOP
           c_in <= '0';
           x <= conv_std_logic_vector(i,k*m);
           y <= conv_std_logic_vector(j,k*m);
           wait for DELAY;
           ASSERT ( z = zz(k*m-1 DOWNTO 0))
           REPORT "error in addition: "
               & integer'image(i) & " + " & integer'image(j) & " + 0"
           SEVERITY ERROR;
           ASSERT ( c_out = zz(k*m))
           REPORT "error in addition: "
               & integer'image(i) & " + " & integer'image(j) & " + 0"
           SEVERITY ERROR;
           c_in <= '1';
           wait for DELAY;
           ASSERT ( z = zz(k*m-1 DOWNTO 0))
           REPORT "error in addition: "
               & integer'image(i) & " + " & integer'image(j) & " + 1"
           SEVERITY ERROR;
           ASSERT ( c_out = zz(k*m))
           REPORT "error in addition: "
               & integer'image(i) & " + " & integer'image(j) & " + 1"
           SEVERITY ERROR;
       END LOOP;
   END LOOP;
   REPORT "simulation OK";
   WAIT;
   END PROCESS;
END test;
```

result is shown in Fig. 8.6 and the following message is sent to the user interface:

** Error: error in c_out: 0 + 63 + 1
Time: 6400 ns

Actually when $x = 0$, $y = 63$ and $c_{in} = 1$, the correct result should be $z = 0$ and $c_{out} = 1$.

However, as shown in Fig. 8.6, $c_{out} = 0$. The simulation execution is aborted.

After correcting the error, the simulation result is shown in Fig. 8.7 and the following message is sent to the user interface:

** Note: simulation OK
Time: 409600 ns

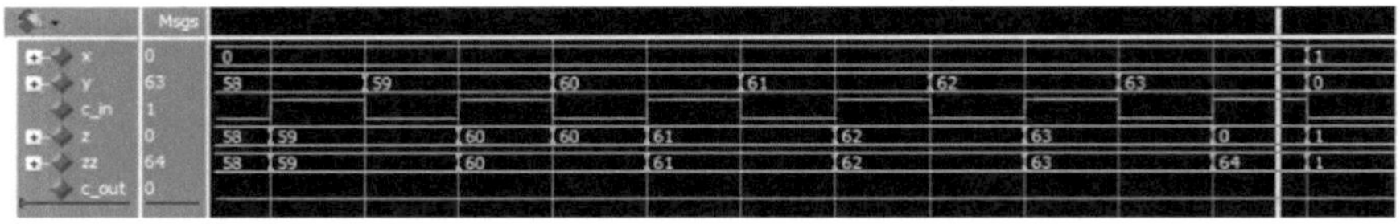

Fig. 8.6 Simulation result (an error is detected) (courtesy of Mentor Graphics)

Fig. 8.7 Simulation result (without error) (courtesy of Mentor Graphics)

8.5.3   Logic Simulation

Hardware description languages, at structural level, are commonly used to describe logic circuits. The instantiated components can be either generic gates, flip-flops, registers, and so on, or elements of a particular implementation library (Standard Cell, Gate Arrays, FPGA, CPLD libraries). Thus, there is no fundamental difference between RT and logic simulation.

components of the UNISIM library for functional simulation of Xilinx primitives, namely flip-flops (FDCE), lookup tables (LUT2, LUT3), an inverter (INV), input and output buffers (IBUF, OBUF) and a clock buffer (BUFGP). The following test bench permits to simulate the synthesized circuit (as far as the UNISIM library is available within the simulation environment).

```
LIBRARY ...
ENTITY test_counter IS END test_counter;
ARCHITECTURE test OF test_counter IS
     COMPONENT counter IS
     PORT (clk, reset, count_enable: IN STD_LOGIC;
           y: OUT STD_LOGIC_VECTOR(2 DOWNTO 0));
     END COMPONENT;
     SIGNAL clk: STD_LOGIC : = '1';
     SIGNAL reset, count_enable: STD_LOGIC;
     SIGNAL y: STD_LOGIC_VECTOR(2 DOWNTO 0);
BEGIN
     dut: counter PORT MAP(clk => clk, reset => reset,
           count_enable => count_enable, y => y);
     clk <= NOT(clk) AFTER 50 NS;
     reset <= '1', '0' AFTER 200 NS;
     count_enable <= '0', '1' AFTER 1500 NS, '0' AFTER 3000 NS;
END test
```

Example 8.9 The RTL definition of a mod 8 counter (*counter.vhd*) has been given in Example 8.1. This RTL description has been synthesized (Fig. 8.2) and the corresponding VHDL model *counter_synthesis.vhd* is available at the Authors' web site. It is a structural description using

This test bench instantiates the device under test and defines the value of all input signals: *clk*, *reset* and *count_enable*. The compilation and execution of this test bench with a VHDL simulator (including UNISIM) permits to visualize the working of the circuit (Fig. 8.8).

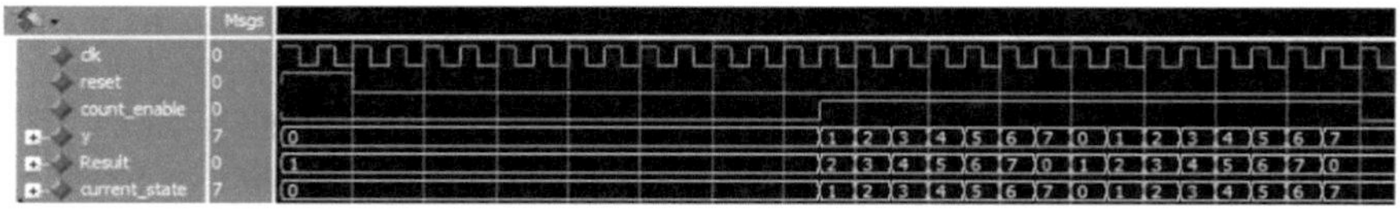

Fig. 8.8 Simulation result (courtesy of Mentor Graphics)

8.5.4 Timing Simulation and Timing Analysis

Once the circuit components have been placed and routed, accurate timing information predictions can be extracted (back annotation, Fig. 8.1) from the implemented circuit description. Before the physical programming (FPGA, CPLD) or manufacturing (SC, GA) of prototypes, those predictions can be used to simulate the circuit with this timing information or to detect and analyze critical paths. This gives the designer the opportunity to detect and to fix possible timing errors before performing costly and irreversible back-end operations.

Back annotation generates HDL files describing the implemented circuit with estimated and accurate internal delays.

Example 8.10 The structural description of a mod 8 counter (*counter_synthesis.vhd*) has been given in Example 8.8. It uses components of the UNISIM library. After implementation and back annotation of the circuit a new VHDL model *counter_timesim.vhd* is generated. It uses components of the SIMPRIM library for timing simulation of Xilinx primitives. This model is available at the Authors' web site. The same test bench as in Example 8.9 can be used (as far as the SIMPRIM library is available within the simulation environment).

Static timing analysis tools compute the expected timing of the circuit without requiring simulation. The word *static* refers to the fact that this analysis is carried out in an input independent manner so that no test bench must be defined. The objective is to find the worst-case delay of the circuit over all possible input combinations. For that the interconnection graph, that represents the final netlist and determines the worst-case delay, is analyzed. The following are some of the circuit characteristics that timing analysis tools compute.

- Critical path: the path with the maximum delay between an input and an output.
- Hold time violation: when an input signal changes too fast, after the clock active transition (race conditions).
- Setup time violation: when a signal arrives too late to a synchronous element and misses the corresponding clock edge (long path fault).

8.6 Other Tools

To conclude this chapter, some other EDA tools are briefly described. They are available within most ASIC and FPGA vendor development packages.

In order to make simpler the design of complex systems, libraries of predefined complex functions and circuits, that have been tested and optimized, have been developed and are available within most commercial EDA tools. They permit to speed up the design process. These predefined circuits are commonly called *intellectual property* (*IP*) *cores* (or *IP blocks*) and are available from semiconductor vendors and third-party IP suppliers. Typical IP cores are arithmetic circuits, memory blocks and memory controllers, clock managers, receivers and transmitters, and others.

Such predefined circuits can be available under the form of RT-level synthesizable hardware description language code (VHDL, Verilog). In some cases, they are offered as generic gate-level netlists. Both netlists and synthesizable cores are called *Soft IP cores*. The logic synthesis and place and route tools treat them as other library components.

Hard IP cores are low-level physical description, for example parts of integrated circuit in transistor circuit format or even in layout format. Thus, they are associated with a particular semiconductor vendor and with a particular technology.

In the case of FPGA chips, heterogeneous blocks such as those mentioned above (multipliers, memories, and so on) can be previously integrated outside the programmable areas. These components are also called Hard IP cores.

Formal verification is a verification tool that, instead of simulating the circuit working with test benches, proves the correctness (or not) of a system using formal mathematical methods. Since hardware complexity growth continuously, the verification complexity is more and more challenging. As a matter of fact, it is impossible to simulate all possible states of a circuit that integrates hundreds of thousands of gates. In order to implement the formal verification, hardware verification languages (HVL) have been defined; they are programming language used to verify the correctness of circuits that are described in a hardware description language (HDL). System-Verilog, OpenVera, and SystemC are examples of commonly used HVL's.

There exist several tools (hardware and software) that permit to make faster the verification tasks and to emulate the working of the circuit under development within its target system. For example:

- *Hardware Emulator*: special purpose hardware that imitates the behavior of the circuit under development.
- *In-circuit Emulator*: hardware that can be plugged into a system in place of the circuit under development (actually a particular case of the preceding).
- *Simulation Accelerator*: a hardware accelerator connected to the used workstation; the accelerator simulates the circuit behavior while the test bench continues to run on the workstation.

Bibliography

Deschamps JP, Sutter G, Cantó E (2012) Guide to FPGA Implementation of Arithmetic Functions. Springer, Dordrecht

Deschamps JP, Valderrama E, Terés Ll (2017) Digital Systems: from Logic Gates to Processors. Springer, New York

Appendix A
Binary Field Operations

Consider the set of binary polynomials of degree smaller than some previously defined constant m:

$$a(x) = a_{m-1}x^{m-1} + a_{m-2}x^{m-2} + \ldots + a_1 x + a_0,$$
$$a_i \in \{0, 1\} \quad \forall i = 0 \text{ to } m - 1.$$

The coefficient operations are the mod 2 sum (XOR function) and the product (AND function). With those operations, the sum and the product of two polynomials can be defined.

A1 Addition

The sum of two polynomials amounts to the mod 2 sum of their coefficients:

$$
\begin{aligned}
& \left(a_{m-1}x^{m-1} + a_{m-2}x^{m-2} + \ldots + a_1 x + a_0\right) \\
& + \left(b_{m-1}x^{m-1} + b_{m-2}x^{m-2} + \ldots + b_1 x + b_0\right) \\
& = (a_{m-1} \oplus b_{m-1})x^{m-1} + (a_{m-2} \oplus b_{m-2})x^{m-2} \\
& + \ldots + (a_1 \oplus b_1)x + (a_0 \oplus b_0).
\end{aligned}
$$

The corresponding computation resource is a set of m XOR2 gates working in parallel (Fig. A.1).

A2 Multiplication

The product of two polynomials is performed modulo a polynomial $f(x)$ of degree m:

$$f(x) = x^m + f_{m-1}x^{m-1} + f_{m-2}x^{m-2} + \ldots + f_1 x + 1.$$

To compute $a(x) \cdot b(x) \bmod f(x)$, first compute $c(x) = a(x) \cdot b(x)$, so that the degree of $c(x)$ is smaller than or equal to $2m - 2$, but could be greater than $m - 1$, and then reduce $c(x) \bmod f(x)$. The definition of the mod $f(x)$ reduction of a polynomial $a(x)$ is similar to the definition of the mod p reduction of a natural number. Given a polynomial $a(x)$, whatever its degree, $a(x) \bmod f(x)$ is the remainder of the division of $a(x)$ by $f(x)$: if $a(x) = f(x) \cdot q(x) + r(x)$, with $degree(r) < degree(f) = m$, then $a(x) \bmod f(x) = r(x)$.

A classical mod $f(x)$ multiplication algorithms use the fact that the mod $f(x)$ multiplication by x is easy:

$$
\begin{aligned}
a(x) \cdot x &= \left(a_{m-1}x^{m-1} + a_{m-2}x^{m-2} + \ldots + a_1 x + a_0\right) \\
& \cdot x = a_{m-1}x^m + a_{m-2}x^{m-1} + \ldots + a_1 x^2 + a_0 x \\
&= f(x) \cdot a_{m-1} - \left(f_{m-1}x^{m-1} + f_{m-2}x^{m-2} + \ldots + f_1 x + 1\right) \\
& \cdot a_{m-1} + a_{m-2}x^{m-1} + \ldots + a_1 x^2 + a_0 x,
\end{aligned}
$$

so that $a(x) \cdot x \bmod f(x) =$

$$
\begin{aligned}
& (a_{m-2} - f_{m-1} \cdot a_{m-1})x^{m-1} + (a_{m-3} - f_{m-2} \cdot a_{m-1})x^{m-2} \\
& + \ldots + (a_0 - f_1 \cdot a_{m-1})x + (0 - a_{m-1}).
\end{aligned}
$$

Thus, as the mod 2 subtraction is equivalent to the mod 2 sum, the coefficient of x^i, with $i > 0$, is equal to

$$a_{i-1} \oplus a_{m-1} \text{ if } f_i = 1 \text{ and to } a_{i-1} \text{ if } f_i = 0,$$

and the degree-0 coefficient is a_{m-1}. The corresponding circuit is shown in Fig. A.2.

The interleaved multiplication algorithm is based on the following equality

© Springer Nature Switzerland AG 2019
J.-P. Deschamps et al., *Complex Digital Circuits*,
https://doi.org/10.1007/978-3-030-12653-7

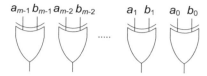

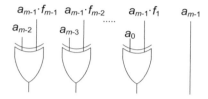

Fig. A.1 Addition of two polynomials

Fig. A.2 Multiplication by x

$$a(x) \cdot \left(b_{m-1}x^{m-1} + b_{m-2}x^{m-2} + \ldots + b_1 x + b_0\right)$$
$$\mod f(x) = a(x) \cdot b_0 + a(x) \cdot x \cdot b_1$$
$$+ a(x) \cdot x^2 \cdot b_2 + \ldots + a(x) \cdot x^{m-1} \cdot b_{m-1} \mod f(x).$$

From the latter expression, the following Algorithm A.1 is deduced. It includes the sum of polynomials, the product of a polynomial by a constant b_i that amounts to multiplying all polynomial coefficients by b_i, and the mod $f(x)$ multiplication by x.

Algorithm A.1 Interleaved multiplication, least significant bit first

```
c(x) = 0;
for i in 0 to m-1 loop
   c(x) = c(x) + a(x)·b_i ;
   a(x) = a(x)·x mod f(x);
end loop;
```

The computation resource that executes the loop body operations is shown in Fig. A.3. Thus, the computation time of a mod $f(x)$ polynomial multiplication is equal to $m \cdot T_{CLK}$ where the clock period T_{CLK} must be greater than the delay of the circuit of Fig. A.3 (two 2-input gate delays).

A3 Squaring

To compute $a(x)^2 \mod f(x)$, the preceding interleaved multiplication algorithm could be used. Nevertheless, taking into account that the operations are executed over a binary field, it can easily be demonstrated that

$$\left(a_{m-1}x^{m-1} + a_{m-2}x^{m-2} + \ldots + a_1 x + a_0\right)^2$$
$$= a_{m-1}x^{2m-2} + a_{m-2}x^{2m-4}$$
$$+ \quad \ldots + a_2 x^4 + a_1 x^2 + a_0.$$

It is a straightforward consequence of the fact that, when multiplying $a(x)$ by itself, partial products such as $a_i x^i \cdot a_j x^j = (a_i \cdot a_j)x^{i+j}$ and $a_j x^j \cdot a_i x^i = (a_j \cdot a_i)x^{i+j}$, with $i \neq j$, cancel each other as $a_i \cdot a_j \oplus a_j \cdot a_i = 0$, while products like $a_i x^i \cdot a_i x^i$ are equal to $(a_i \cdot a_i)x^{2i} = a_i x^{2i}$.

Assume now that a set of binary constants r_{ij} has been previously computed. They are defined by the following relations:

$$x^{m+i} \mod f(x) = r_{m-1,i}x^{m-1} + r_{m-2,i}x^{m-2}$$
$$+ \ldots + r_{1,i}x + r_{0,i}, i = 0 \text{ to } m - 2.$$

Assume that m is odd and equal to $2\,k-1$. Then,

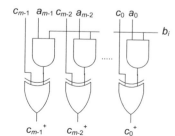

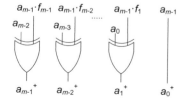

Fig. A.3 Interleaved multiplication step

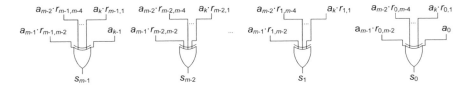

Fig. A.4 Squaring

$$a_{m-1}x^{2m-2} + a_{m-2}x^{2m-4}$$
$$+ \cdots + a_2x^4 + a_1x^2 + a_0$$
$$= a_{m-1} \cdot \left(r_{m-1,m-2}x^{m-1} + r_{m-2,m-2}x^{m-2} \right.$$
$$\left. + \cdots + r_{1,m-2}x + r_{0,m-2} \right)$$
$$+ a_{m-2} \cdot \left(r_{m-1,m-4}x^{m-1} + r_{m-2,m-4}x^{m-2} \right.$$
$$\left. + \cdots + r_{1,m-4}x + r_{0,m-4} \right)$$
$$+ a_k \cdot \left(r_{m-1,1}x^{m-1} + r_{m-2,1}x^{m-2} \right.$$
$$\left. + \cdots + r_{1,1}x + r_{0,1} \right.$$
$$+ a_{k-1}x^{m-1} + a_{k-2}x^{m-3}$$
$$+ \cdots + a_1x^2 + a_0.$$

Finally, $a(x)^2 \bmod f(x) = s_{m-1}x^{m-1} + s_{m-2}x^{m-2} + s_{m-3}x^{m-3} + \ldots + s_1x + s_0$ where

$$s_{m-1} = a_{m-1} \cdot r_{m-1,m-2} + a_{m-2} \cdot r_{m-1,m-4}$$
$$+ \ldots + a_k \cdot r_{m-1,1} + a_{k-1},$$
$$s_{m-2} = a_{m-1} \cdot r_{m-2,m-2} + a_{m-2} \cdot r_{m-2,m-4}$$
$$+ \ldots + a_k \cdot r_{m-2,1},$$
$$s_{m-3} = a_{m-1} \cdot r_{m-3,m-2} + a_{m-2} \cdot r_{m-3,m-4}$$
$$+ \ldots + a_k \cdot r_{m-3,1} + a_{k-2},$$
$$\cdots$$
$$s_1 = a_{m-1} \cdot r_{1,m-2} + a_{m-2} \cdot r_{1,m-4}$$
$$+ \ldots + a_k \cdot r_{1,1},$$
$$s_0 = a_{m-1} \cdot r_{0,m-2} + a_{m-2} \cdot r_{0,m-4}$$
$$+ \ldots + a_k \cdot r_{0,1} + a_0.$$

Thus, every coefficient s_i of $a(x)^2 \bmod f(x)$ is the sum of at most $k = \lfloor m/2 \rfloor$ coefficients a_j. In the case of the most commonly used polynomials $f(x)$, the matrix $[r_{i,j}]$ has few nonzero coefficients so that every coefficient s_i is the sum (XOR gate) of a few coefficients a_j (Fig. A.4).

Another interesting property of binary fields (more generally, binary commutative rings) is that

$$(a(x) + b(x))^2 = a(x)^2 + a(x) \cdot b(x)$$
$$+ b(x) \cdot a(x) + b(x)^2 = a(x)^2 + b(x)^2,$$
$$(a(x) + b(x))^4 = \left((a(x) + b(x))^2 \right)^2$$
$$= \left(a(x)^2 + b(x)^2 \right)^2 = a(x)^4 + b(x)^4,$$

$$(A.1)$$

and so on.

Appendix B
Elliptic Curves

In this appendix, a particular type of elliptic curve, namely Koblitz curves over a binary field $F = GF(2^m)$, is defined. The binary field F (Appendix A) consists of all binary polynomials of degree smaller than m, with operations modulo an irreducible (non-factorizable) polynomial $f(z)$ of degree m:

$$f(z) = z^m + z^{m-1} + \ldots + z + 1.$$

B.1 Definition

An elliptic curve $E(F)$ is defined as follows: it consists of all pairs $(x, y) \in F^2$ of binary polynomials such that $y^2 + xy = x^3 + x^2 + 1$, plus a particular element ∞ called *element at infinity*:

$$E(F) = \{(x,y) \in F^2 | y^2 + xy = x^3 + x^2 + 1\}$$
$$\cup \{\infty\}.$$

$$(B.1)$$

It has been demonstrated (Hasse theorem) that, for great values of $q = 2^m$, the number of points of $E(F)$ is approximately equal to the number of field elements:

$$\#E(F) \cong q = 2^m. \qquad (B.2)$$

B.2 Additive Group

An addition operation can be defined over $E(F)$. The so-obtained algebraic structure $(F, +, \infty)$ is a commutative group. The addition is defined by the following rules.

(1) ∞ is the neutral element: $P + \infty = \infty + P = P$, $\forall P \in E(F)$.

(2) The inverse of $P = (x, y)$ is $-P = (x, x + y)$.

(3) If $P = (x_1, y_1)$, $Q = (x_2, y_2)$, $P \neq Q$ and $P \neq -Q$, then $P + Q = (x_3, y_3)$ where

$$x_3 = \lambda^2 + \lambda + x_1 + x_2 + a,$$
$$y_3 = \lambda(x_1 + x_3) + x_3 + y_1,$$
$$\lambda = (y_1 + y_2)/(x_1 + x_2).$$

(4) If $P = (x_1, y_1)$ and $x_1 \neq 0$, so that $P \neq -P$, then $P + P = (x_3, y_3)$ where

$$x_3 = \lambda^2 + \lambda + a = x_1^2 + b/x_1^2,$$
$$y_3 = x_1^2 + \lambda x_3 + x_3, \lambda = x_1 + y_1/x_1.$$

According to (B.2), the order of this group (the number of elements) is approximatively equal to 2^m.

B.3 Scalar Product

Given a point P of $E(F)$ and a natural k, the scalar product kP is defined by

$$kP = P + P + \ldots + P \ (k \ \text{times}),$$
$$\forall k > 0 \ \text{and} \ 0P = \infty.$$

The order of a particular element P of $E(F)$ is the smallest integer s such that $sP = \infty$ (the neutral

© Springer Nature Switzerland AG 2019
J.-P. Deschamps et al., *Complex Digital Circuits*,
https://doi.org/10.1007/978-3-030-12653-7

element). A well-known property of finite commutative groups is that the order of an element divides the number of elements of the group.

Some elliptic curves (E_1-type Koblitz curves) have the following property: the number of elements of the associated group $E(F)$ is equal to $2n$ where n is a prime. Thus, as the order of an element P divides $2n$, the order of P is 1, 2 or n. If the order of P is equal to 1, then $P = 1P = \infty$. If the order of P is equal to 2, then $P \neq \infty$, $P = (x, y)$, $(x, y) + (x, y) = \infty$, $(x, y) = -(x, y) = (x, x + y)$ and $x = 0$. Thus, if P is neither ∞ nor $(0, y)$, then its order is equal to n, where $n = \#E(F)/2 \cong 2^{m-1}$ and all the scalar products kP, with $k \in \{0, 1, 2, \cdots, n-1\}$, have different values.

B.4   An Example of Koblitz Curve Over a Binary Field

Consider the elliptic curve (B.1) where $F = GF$ (2^{163}) and $f(z) = z^{163} + z^7 + z^6 + z^3 + 1$. The order of the following point $P = (x_P, y_P)$ where x_P and y_P are binary polynomials represented as 41-digit hexadecimal numbers smaller than $2^{163} = 8 \cdot 16^{40}$

$x_P = $ 2fe13c0537bbc11acaa07d793de4e6d5e5c94eee8,

$y_P = $ 289070fb05d38ff58321f2e800536d538ccdaa3d9,

is equal to

$n = $ 4000000000000000000020108a2e0cc0d99f8a5ef

(a 41-digit hexadecimal number). Observe that $n \cong 4 \cdot 16^{40} = 2^{162}$.

The function $h: \{0, 1, 2, \cdots, n - 1\} \to E(F)$ defined by $h(k) = kP$ is a one-way function. That means that even if the values of P and $h(k) = kP$ are known, it is very difficult to compute the value of k. Actually, there is no simple algorithm to compute h^{-1}, and the trivial algorithm that consists in computing sP for all s in $\{0, 1, 2, \cdots, n - 1\}$ until $sP = kP$ (and thus $k = s$) has a practically infinite computation time: assume that a scalar product can be executed in 1 microsecond; to compute kP for all k in $\{0, 1, 2, \cdots, n - 1\}$ with $n \cong 2^{162}$, the computation time is about 2^{162} s and is longer than 10^{34} years $(2^{162} > 10^{48}/366 \cdot 24 \cdot 3600 > 10^{34})$.

Index

© Springer Nature Switzerland AG 2019
J.-P. Deschamps et al., *Complex Digital Circuits*,
https://doi.org/10.1007/978-3-030-12653-7

Printed in the United States
By Bookmasters